The
Trump Century

ALSO BY LOU DOBBS

*Upheaval: Winning Back the Country with Knowledge That
Empowers, Ideas That Matter, and Solutions That Work*

Independents Day: Awakening the American Spirit

*War on the Middle Class: How the Government,
Big Business, and Special Interest Groups Are Waging
War on the American Dream and How to Fight Back*

WITH JAMES O. BORN

Border War

Putin's Gambit

The Trump Century

How the President Changed
the Course of History Forever

LOU DOBBS

with DENNIS KNEALE

BROADSIDE BOOKS
An Imprint of HarperCollins*Publishers*

HarperCollins books may be purchased for educational, business, or sales promotional use. For information, please email the Special Markets Department at SPsales@harpercollins.com.

Broadside Books™ and the Broadside logo are trademarks of HarperCollins Publishers.

FIRST EDITION

Library of Congress Cataloging-in-Publication Data has been applied for.

ISBN 978-0-06-302904-0

20 21 22 23 24 LSC 10 9 8 7 6 5 4 3 2 1

For all our patriots of every kind and creed
who make America great and lasting

We frequently see the respectful attentions of the world more strongly directed towards the rich and the great, than towards the wise and virtuous. We see frequently the vices and the follies of the powerful much less despised than the poverty and weakness of the innocent.

—*Adam Smith*

CONTENTS

INTRODUCTION

We are living through the damnedest, most chaotic times of our lives. Only months ago we were enjoying boom times in the Trump era, and the next moment we were fighting a mysterious new virus in one of the worst crises in our nation's history. This stunning reversal occurred with only months to go in the 2020 presidential election, putting at risk not just the next four years but perhaps the next twenty or thirty years and beyond.

As I write these words to you on Memorial Day 2020, from our farmhouse some miles from Sussex, N.J., all of America is on lockdown. So am I. In self-quarantine for two months now, I am fortunate to be able to host *Lou Dobbs Tonight* on the Fox Business Network from my home.

We put the show in quarantine on Friday, March 13. I had visited Vice President Mike Pence in Washington to interview him for my program on March 4. One of my staffers later became Patient Zero at the Fox News building on the Avenue of the Americas in midtown Manhattan. He is fine. A week later we were up and running from our makeshift studio, equipped with a remote camera and lighting and a hacker-proof digital fiber line, with a satellite truck parked in my driveway.

For the first three furious and frantic years of the presidency of Donald Trump, I have had a front row seat. Now and then I talked to the president directly, usually in private phone calls early in the morning or late in the evening, discussing the economy, foreign policy and other matters at hand in his bruising battles with so many enemies.

I watched with admiration and fascination as he toppled the old order and defeated fiery foes to pursue his America First agenda. In his first three years in office, the most successful of any presidency, he had achieved a litany of wins that was nothing short of amazing, on so many fronts that were of vital importance to our country: jobs, growth, immigration, the China threat, and so much more.

My delight turned to concern and a new focus on crisis management amid the devastation that China had unleashed upon America and the world. For twenty years China crafted the image as an economic miracle that lifted a billion people out of poverty. Now, China's real legacy is the Wuhan virus and an odious proclivity for delaying, dissembling, and lying to protect its own reputation no matter what the damage has done to everyone else.

This book is about what happened when the irresistible force of President Trump slammed into the most immovable object of all time and smashed it to pieces. His cosmic collision with the fevered opposition created one of the most impossible, unbelievable, riveting times in our history. At stake was Trump's bid to alter the future course of America and end thirty years of bad trade and weak borders.

The immovable object blocking President Trump's path arose from thirty years of free-trade orthodoxy, weak industrial policy, and promiscuous practices in immigration. Complicit in this downward spiral: neglectful presidents, arrogant advisors, self-serving incumbents in Congress, slack regulators, insolent bureaucrats, and liberal judges.

They were joined by Big Business, Big Labor, dishonest American economists on China's payroll, and NGOs masked as nonpartisan entities. And cheered on by the left-wing media.

The throngs of detractors failed to realize that most everything that was happening, even the setbacks that jolted the new administration, resulted from the strategy and tactics shaped by President Trump and his team.

Hollywood would reject this pitch out of hand as too far-fetched—even if it were about a liberal. Billionaire turned populist, vivid and larger than life, in a power struggle for the soul and the destiny of our country. Born into wealth and a builder of luxury hotels with gold-plated faucets, he gets elected by championing laid-off factory workers of the Rust Belt, who were abandoned by Washington long ago. The phrase itself dates to a cover headline in *BusinessWeek* magazine—in 1980.

Most of those who preceded him, Democrat or Republican, had climbed the ranks of local politics, state legislatures, the governorship, or a congressional seat. Donald Trump had no political experience at all. He came from a glittering penthouse in Trump Tower in midtown Manhattan, descending the escalator into the lobby to tackle some of the greatest challenges we ever had faced.

He took on everyone—all comers—as he took aim at the ossified, petrified orthodoxy across the political spectrum. The ideological fools fighting his every effort had no sense of the causal relationship between policy and consequences, whether in weak foreign policy or the moribund economy over which Barack Obama presided for eight years. They had no more regard for the policies of George W. Bush than they had for those of Bill Clinton. They benefited just fine under both.

President Trump did this by disrupting the old dogma on free trade that had enriched the Washington establishment and the globalist

elites for decades. The powerful populist from Queens, N.Y., wanted
to end America's role as the grand chump on the globalist One World
stage. My gawd, he dared to put America first! And he pursued this
agenda with an abundance of original, brighter, and bolder ideas. The
globalist elites fought him furiously not because they believed he was
evil, but because his rip-snorting patriotic agenda posed a risk to their
wealth.

They wanted him gone. So badly that the Deep State waged a
years-long, invisible coup attempt to remove President Trump from
office. This scandalous and infuriating plot was aided and abet-
ted and covered up by the Fake News left-wing corporate media—
co-conspirators in this sleazy assault. In the Trump era, the Fourth
Estate became the Fifth Column. The media joined the Radical Resistance's destructive efforts to remove a newly elected president from
office even before he could get started.

The scheme was criminal, unconstitutional, and the biggest threat
to American democracy and the presidency in our nation's history.

Most despicably of all, it was waged secretly by the highest levels
of the agencies we need to trust the most: the FBI, the CIA, and the
NSA. They used bogus intelligence, paid for by the Hillary Clinton
campaign, to lie to the foreign-surveillance court to obtain warrants
to spy on Trump officials. They were assisted by the Democrats, radical left social justice warriors, foreign spy agencies—and leaders of
the Republican Party, who failed to fight back and come to President
Trump's full-throated defense.

Yet he overcame them all. The worst of these offenders deserve
harsh punishment, and the American people must demand it no matter what the outcome of the election. As President Trump handled the
Wuhan crisis, the polls were giving his likely opponent, Democrat
Joe Biden, a 10-point lead or better. It was unclear whether Donald

Trump, after racking up so many impossible victories in his first three years in the White House, was going to win this one.

No matter what the outcome of the 2020 election, however, it now is clear that President Trump's impact on America will compound and reverberate for decades to come. As he logged win after win, he changed the conversation on our most important and combustible issues. America will never be the same because of him—it will be better.

Thus, the title of this tome: *The Trump Century*.

This is my front-row take on the Trump presidency and how he managed to rack up a remarkable record of victories and overcome the Resistance, which did all it could to remove him from office.

It is almost unimaginable that President Trump could have survived so many threats to his presidency, from so many quarters, much less thrived the way he did on multiple fronts. What a wild, wild ride it was for all of us, and especially for him. President Trump is the most misunderstood, mischaracterized, and unfairly demonized POTUS ever to occupy the office. He also is the greatest president of the United States of all time. The GOAT among POTUS greats.

Seven American presidents have served in my years as a journalist and then as a TV host (these are two different species). I see President Trump as their equal or better in terms of his acuity, judgment, and, most of all, his zeal and capacity for conflict. He has earned his way into the presidential pantheon of our greatest leaders: Washington, Jefferson, Lincoln, Wilson, FDR, Truman, Kennedy, Reagan.

None of them had to endure the hatred, disrespect, and false accusations thrown at President Trump. There isn't another person in the country who could have accomplished anywhere near what President Trump did in three years. And no one else could have stood up to the constant, unrelenting venal attacks he withstood with such courage and strength.

In terms of a single defining president who held great sway and set the tone for an entire century, one could say it was Lincoln in the 19th century, and for the 20th century, Reagan or FDR. Now comes the 21st century, and the Trump legacy will be inescapable.

The Wuhan virus crisis had been so all-consuming that it diverted America's attention from Deep State coup attempt, which had begun even before President Trump was elected, mutating and morphing and growing more menacing by the time of Wuhan. This was an outrage, a scandal on a monumental scale that dwarfs the worst setbacks in our history. Watergate, the first of many "gates" to spill out over the next five decades, was a fraternity prank by comparison.

The FBI later admitted this. It cited "errors" in the FISA applications. Errors that always went in favor of spying on Trump associates so they could, in essence, spy on the president they wanted to remove from office.

President Obama, so revered by the media as beyond all reproach and any suspicion, personally approved the effort. Two weeks before leaving office, in a group meeting in person, he told the FBI to keep him apprised of the investigation. We know this because Obama loyalist Susan Rice wrote an email to herself in the moments after President Trump had just been sworn into office. She was trying to cover for Obama, and, instead, she had looped him in and made it clear he had known of this plot for some time.

Vice President Joe Biden was in on it, too. He was one of three dozen different Obama officials to request NSA permission to unmask the identity of Gen. Flynn in the Russian call; this would let them leak Flynn's name and the scandal to the left-wing media. Of course, the Flynn name was leaked regardless, that is how the FBI plays this game.

FBI Director James Comey later bragged, in an onstage interview to an admiring, chuckling crowd, about sending two agents over to interview Gen. Mike Flynn at the White House in the chaos of the first few weeks of the Trump administration. They told Flynn he had no need for a lawyer's counsel, this was routine. They had listened in on Flynn's call with a Russian official and had a transcript of it, and they asked him about it to entrap him for a federal perjury charge, purposely and premeditatedly.

When Flynn stood strong, they threatened to take down his son. They were brutal, merciless, immoral, criminal—and FBI Director Comey, later fired by President Trump, turned this outrage into a personal publicity tour. As did James Clapper, who was Obama's director of national intelligence, and former CIA director John Brennan, who were deeply complicit and helped lead this coup attempt against a sitting president. The federal penalties for treason include the death penalty by hanging; given the extent of the crimes committed in the Russiagate scandal, one can see why that is so.

President Trump overcame even this attack on his presidency, a feat so historic and courageous that, one day, long after his liberal haters fade from existence, historians will marvel at the miracles he achieved in the face of the ugliest and most vehement and criminal opposition any president has ever had to conquer—especially in his first term in office. There are some days I cannot believe it, that his angriest, most desperate and most vicious enemies refuse to credit him for his many monumental successes.

In the incessant media storm that attempted to distort every aspect of Trump's life and his actions and words, someone had to tell the real story of this president, and provide an honest and true assessment illuminating his character and qualities, all that he stands for and all that he has achieved. Someone like yours truly, an observer of presidents and their policies, who agrees with this president on almost every issue.

I started writing this book at the height of the Trump economic boom and soaring, record-high markets. My thesis is straightforward: President Trump has been right, and his opponents very wrong. He won, and they lost. Elections have consequences, as his predecessor said. Yet they are welcome to enjoy the benefits of the brighter and better future he is building for all Americans—and stop trying to block every single thing he proposes. That, or every one of the members of the resistance can go straight to hell; that's always an option.

The consequences of Trump's election were apparent in record time. The wags on Wall Street and the national media insisted the Trump agenda would amount to more Obama-like policy drift and devastate our economy and market and accelerate the decline of the American worker and entrepreneur.

My goal is to provide evidence and solid arguments to help President Trump win re-election. Doing so is even more important now. After Wuhan, America needs a builder, now more than ever before.

Throughout the Trump presidency, his unyielding and venal opponents have outnumbered him and done everything in their power, whether on the part of the radical Dems or the Deep State, to destroy President Trump and his agenda. His driving motivation, by contrast, has been America first, expressed by the red baseball cap with the president's unassailable slogan and strategic goal, Make America Great Again. President Trump is the nation's First Patriot, and in all things, he puts America first, and that is all he asks of every American.

The Radical Left, by contrast, avoided debating the facts and, instead, boiled everything down to charges that President Trump was racist. Or xenophobic. Or authoritarian. They refused any solutions that entailed working with him. They ended up settling for nothing. They were watching a true leader lead from the front, rather than from behind as in the Obama era.

Meanwhile, the president's style as a ferocious counterpuncher was another of his many strengths. When the haters attacked him with vituperative venom, he doubled down and slapped them back. The clashes were epic, and this made Trump's victories all the more remarkable—and all the more delightful, inspiring even more fervent devotion from his fans.

Because of his life experiences, Donald Trump played by a different set of rules that were alien to politicians, policymakers, interest groups, and the Washington and corporate ecosystem that feeds them. He spoke casually, without a filter, when other pols were deceitful, calculated, and careful in measuring their every word. It was a refreshing departure from eight years of President Obama's halting style, with an abundance of "nots" and double-negatives and liberal doses of "I" and "me." Trump prefers "we" and "us," and he sings the praises of those who help him rebuild America.

President Trump lacked government conditioning on the way things worked. He had no patience for long-standing protocols and taboos that were out of bounds for a U.S. president to broach. This became another strength. It freed him to address issues he was supposed to avoid, and to do things that were undoable.

Things like imposing tariffs on China, berating the Fed for raising interest rates, calling out our allies in NATO for welching on defense, cutting back on the number of refugees who are granted asylum in the U.S., and erecting a wall along parts of the border to impede the flow of illegal aliens, drugs, and criminals. He was bluntly honest, he told the unvarnished truth, and his supporters loved him for it.

He also scrapped the hallmark deals that comprised the Obama legacy: the Trans-Pacific Partnership, Paris climate pact, and the Iran nuclear deal; as well as NAFTA, passed in the Clinton era and hailed by Bill Clinton as a job creator. It was. For Mexico.

President Trump wasn't supposed to do that, yet he did all of it.

This benefited America, hugely. I am unable to name a single instance in which the president guessed wrong and America suffered the dire consequences and doomsday effects predicted so often by his haters.

He also has transformed the Republican Party and pushed the divisive issues of trade policy and immigration squarely to the center stage of American life and politics. That will remain true long after President Trump leaves Washington. His successor will be proceeding in the tracks laid by President Trump, and building on the profound and fundamental changes he put into place.

Now that China has bowed to the Trump tariffs and made major concessions on trade, no future president dare peel that back at the expense of American workers. No U.S. multinational ever again will blithely shutter factories and ship work overseas without harder consideration of the real implications and costs.

Most of us would be beleaguered by the mean-spirited opposition that besieged President Trump for his entire presidency. He thrived on it. The man was inexhaustible. In our off-the-record phone calls, he was the same person at 7:30 in the morning as he was at 10:30 at night, invigorated and high-energy. I am the same person, too, at 7:30 a.m. and 10:30 p.m., but by 10:30 I am *tired*. And I sound like it.

It is common among those who love our president to marvel at his stamina and grit, at how many times he gets knocked down and springs back up swinging, ready for another round. Many of us might avoid the hassle and buy our own island in the Caribbean if we had the wealth Donald Trump possesses.

This book went to press as our president was pushing to reopen America and getting static from Democratic governors who wanted to stay closed longer—till after the election, of course. I had hoped we might see Trump Hater Fatigue by now, that the frantic and fearmongering crusaders of the left would tucker out. That was not to be.

Meanwhile, President Trump has mellowed a bit. He meanders less often into accidental volleys, and he has softened his wording and tempered some aspects of his approach. He avoids roasting someone without provocation on the social media platform he prefers, Twitter, a critical tool for bypassing the Fake News cabal.

For historians, presidential scholars, and political scientists, assessing the impact of President Trump and his agenda will be like evaluating a fine wine. It will get better with each passing decade. Given the liberal bent of the media and academia, it may take them forty or fifty years to gain full appreciation of the boldness and subtleties of the Trump era. By reading *The Trump Century*, you can do that right here, right now.

One truth has been proven to be undeniable. One person, the right person at the right time, can make all the difference in the world. The president said this once, in an interview forty years ago. America was fortunate that our person turned out to be Donald Trump. After Wuhan, we are even luckier that his time is now.

The
Trump Century

CHAPTER ONE

AMERICA FIRST

D onald John Trump has beaten back the haters and emerged from a cauldron of conflict to become the greatest president in America's history. He has joined the pantheon of presidents, including Thomas Jefferson, Abraham Lincoln, and Franklin Roosevelt, who have set the agenda for the decades that followed, guiding our nation through some of its hardest times while shaping the future before it arrives. He has delivered on the unassailable slogan that drove his first campaign for president:

Make America great again.

President Trump has overcome legions of enemies and racked up a string of seemingly impossible victories: stoking robust economic growth, lifting wages for the lowest-paid workers, delivering record-low unemployment, bringing manufacturing jobs home to America, forcing China to the table to reform unfair trade terms, and still more.

And then it all came tumbling down, toppled by an invisible enemy, the biggest threat ever to confront the United States. Now we face a challenge like none before. China's problems became everyone's problems as the Wuhan virus crisis cratered the world economy and wrought devastation in every nation that shut down in response to it. As this book goes to press, thousands of people are still dying and millions are being laid off from their jobs. The only certainty is the expanding regime of profound restrictions on our society, which violate fundamental freedoms of the Constitution.

This tests the very core of what makes the United States the freest, richest, most innovative, and most inclusive nation on Earth. What we do now, and how we go about pulling our way up out of this shockingly deep downturn, could determine our future for decades to come. We can arise from this global pandemic better and stronger than ever before, reigniting growth and fixing the weaknesses that have been revealed. Or we can barely crawl back to a cowering *"new* new normal"* of retreat, fear, recrimination, and diminished freedom masked as protective measures imposed on us for our own good.

Whether we thrive or barely survive will rely in part on who wins the 2020 presidential election: Donald Trump or Joe Biden.

President Trump, I truly believe, is the only leader who has shown that he is bright enough, strong enough, shrewd enough, and original enough in his thinking to lead the United States out of this maelstrom. He is the essence of the right man in the right place at the right time, all for the good of our country. He is an irrepressible salesman who sets impossibly high goals so that even if he gets only halfway there, he ends up twice as far ahead as he would have otherwise. His unrelenting optimism and his skills as a dealmaker, schmoozer, and obstinate negotiator are mission-critical to the gigantic task that awaits us:

Make America *grow* again.

If he can win reelection to a second term, he will get the chance to score his biggest win of all: guiding us to a robust recovery after the steepest, fastest collapse in history—a Great Cessation, some call it. If President Trump can win again, he would join the pantheon of

Winston Churchill and Alexander the Great among the great leaders of world history. And if he falls short, well, there is no point in considering the possibility. He's a winner—*the* winner.

The Wuhan pandemic, China's culpable role in it, and our recovery from it make up the biggest story of my career. It dwarfs all else. China authored the Wuhan virus, whether accidentally or intentionally, and spread it to an unsuspecting world instead of warning anyone. China created the worst crisis for the United States since World War II and is responsible for the deadliest attack on the United States in its history.

The Wuhan shock culminated twenty years of China's ripping off the United States and draining away almost 4 million US manufacturing jobs. The Chinese reaped more than a trillion dollars on unfair trade, currency manipulation, brutal protectionism, extortion of US multinationals, and forbidden government subsidies of home industries.

President Trump's forcing China to sign a new trade deal fixing those ills was one of the most important wins of his presidency. If we lose President Trump, what happens to that historic deal? Or what happens when China tries to push our growing US Navy fleet out of the South China Sea? Or when the president follows through on his threat to withdraw the United States from the China-cowed World Health Organization and form a new global group that includes Taiwan, a thorn in China's side for a hundred years?

We should also ponder the decisions that President Trump or a weaker president would face if the Chinese government were exposed as having engineered the virus in the two government labs in the city where the pandemic first erupted.

Our nation will spend $10 trillion to rebuild from the ashes of the coronavirus. That is equivalent to half the entire output of the US economy in a year. Who best to ride herd on this gargantuan spending project for the next four years: the president who led us to one of the strongest economies of all time or Joe Biden?

Barack Obama and Joe Biden, from the end of the Great Recession in June 2009 till they left office at the end of 2016, presided over one of the pokiest recoveries ever. Growth was slack when it should have been surging, although the United States did log the longest economic expansion ever, and it continued in Trump's first three years.

For the legions of Trump haters who voted for Joe Biden, the bad news is this: the impact of President Trump *is never going away*. The changes he has engineered in our nation's fabric will reverberate for decades. In the long term, it doesn't matter what happens in 2020, 2024, or even 2040.

The resistance has yet to realize the truth: we now live in the Trump Century. President Trump has changed everything. In government and politics, international relations, global trade, business and regulation, the national conversation, and our messaging, social media, branding, and verbal combat.

Many of us are unaware of this reality, especially the elites who despise him most. For them, this moment is akin to the opening scene in *The Matrix*, when the hero, Neo, swallows the red pill (a fitting color) and learns his reality has been all illusion, a virtual reality simulation, and now he must see the truth.

The president's endless enemies hate this, yet it is true. Never Trumpers, Democratic leaders, RINOs (Republicans in Name Only), the Squad, the Swamp, the Deep State, ultraliberal Hollywood, the left-wing corporate media led by MSNBC, CNN, the *New York Times* and the *Washington Post*, and the liberal billionaires of Silicon Valley: a truly formidable phalanx of foes.

The Fake News media and liberal apologists ascribe the fierce resentment of President Trump to some sort of denial and psychosis, a bogus diagnosis of Trump Derangement Syndrome. This is apologist bunkum. The real reason so many globalist elites fought our president so vehemently was that his agenda threatened the very things that made them elite—and wealthy.

The issues we will debate for two or three generations will be the issues President Trump pushed from the periphery to the center stage of US politics and policy. The Democrats will win battles, but only within the framework and context of the Trump agenda. His lifelong ethos still will be shaping the platform of some socialist college freshman of today who gets elected president forty years from now.

The president's platform, succinctly stated: America first. Strong borders, controlled legal immigration, and a strong military. *Fair* trade rather than the lie of "free" trade. A crackdown on China, a tougher tack on Iran, and an antipathy to involving the United States in foreign wars. Plus a harsh and publicly reinforced view of US companies that for years shipped jobs overseas while laying off workers here.

The elites did all they could to thwart the Trump agenda and remove him from office. President Trump outsmarted them all. If those venal, vicious opponents had held their fire and given the president a chance at the start—if they had rooted for his success on America's behalf instead of mobilizing to block his every effort—his administration could have achieved even more. For all of us.

President Trump was out to destroy the antiquated, dishonest orthodoxy on the free-trade policies and globalism that have wreaked destruction on US workers and the US economy for thirty years. Anti-US policies acquiescing to unfair trade terms, offshoring US jobs, weak borders, and unrestrained immigration gave way to the new regime of Trumponomics.

That posed an existential threat to the disciples who had been advocating the damaging policies of global trade: big business, government, the leaders of both political parties, academia, and economists. Only a handful of honest economists exist in this country, and not a damn one of them works for a major US corporation.

This malevolent orthodoxy is coming undone. The only thing holding it together is the massive power of corporate America. It will take another ten years to dissolve, but we are seeing its undoing now. The tired old argument of the globalist elites is that US companies can grow their profits better by shipping work to cheaper labor markets overseas and laying off their workers here at home.

That was always tragically shortsighted and un-American. No one in power did a thing to correct it; not the Democrats, who were natural enemies of big business, nor the Republicans, who were allied with it. The cheap-labor arbitrage sacrificed the purchasing power of workers and suppressed wage growth. Back at home, wages also were undercut by the rampant immigration of illegal aliens willing to work for very low wages under the table in gritty, dirty jobs.

This hollowed out US manufacturing so badly that we are reliant on other nations to produce the things we no longer make. It is an embarrassment.

Yet the US Chamber of Commerce preached the gospel of free trade, as did the Business Roundtable and Wall Street CEOs. Labor unions, once a countervailing force, have been coopted. They now support open borders and the export of US jobs. In fact, unions represent less than 7 percent of the private workforce—and 40 percent of government workers.

The American middle class lacked any power to stop the damage inflicted by our trade policies. Until President Trump came along.

My unalloyed admiration for him owes in part to the fact that, like the president, I have spent more than twenty years haranguing politicians, policy makers, viewers, and readers about the damage our trade policies were inflicting on the United States. My message has been steadfast, whether on my nightly programs on cable news, first on CNN and later on Fox Business Network, in written op-eds, or in testifying before Congress in 2007.

On March 28, 2007, I testified before a House of Representatives subcommittee as Congress was considering reforms that would grant amnesty to millions of illegal residents. In my statement, "There's Nothing Free About Free Trade," I cited numbers that were stunning at the time, and they are even worse a decade later:

The United States has sustained 31 consecutive years of trade deficits, and those deficits have reached successively higher records in each of the past five years. . . . Since 1994, the first full year in which the North American Free Trade Agreement was in effect, the United States has accumulated more than $5 trillion in external or trade debt. . . .

There is nothing free about ever-larger trade deficits, mounting trade debts and the loss of millions of good-paying American jobs. Since the beginning of this new century, the United States has lost more than three million manufacturing jobs . . . to cheap overseas labor markets as corporate America campaigns relentlessly for "higher productivity," "efficiency," and "competitiveness," all of which have been revealed to be nothing more than code words for the cheapest possible labor in the world. . . .

Because I seek balance and reciprocity in our trade policies, I've been called a "table-thumping protectionist," and the Bush administration has hurled at me its favorite public epithet, "economic isolationist." Nothing could be further from the truth. I believe, as I hope you and the majority of all members of this Congress believe . . . in the importance of an international system of trade and finance that is orderly, predictive, well-regulated, mutual, and fair.

It would take a full decade from that point until, at long last, an American president shared my sentiments and urgency and felt driven and duty bound to do something about it. Thank you, Donald Trump.

The first three years of the Trump administration were probably the most successful of any presidency in US history. It was a miraculous and marvelous journey, marred by moments of panic, hysteria, and raw hatred in the opposition. President Trump had braced for this upon taking office.

He said as much in a long conversation with me on my nightly program on Fox Business in October 2017, only nine months into his presidency. His comments were prescient and uncanny in anticipating the unfolding events of the ensuing two years. I had traveled to Washington to interview him in the Cabinet Room of the White House. Our conversation took place on the day of October 25, 2017, and aired that evening.

At the time, the president was pushing Congress to pass the dramatic tax cuts that were the cornerstone of his economic revival policies and embarking on an ambitious series of other forays. We opened the show with a live intro that captured Trump at the center of the storm:

LOU DOBBS: Good evening everybody. We're coming to you from the swamp, Washington, DC, and I can tell you, we come here rarely and only for the most important of reasons. And that is to speak today with President Trump. President Trump promised voters to drain the swamp, but he's also finding the swamp filled with creatures who are more cunning, more elusive, and more dangerous than even he expected when he was campaigning for the job he now holds.

From there, our interview felt like a lunchtime chat between two old friends who were discussing politics. Moments later, the president held forth:

PRESIDENT TRUMP: We are going to have so many jobs created by what we're doing and, you know, we're going to have companies that aren't leaving our country any more. You look at what's happened over the past 25 years, you go down and take a look at Mexico and see how many companies have built these massive factories in Mexico, not here.

We want them built in Michigan and Ohio and our places, and they just leave, and they fire everybody and they go and open plants [in Mexico]. That's not going to be happening any more, plus they are penalized in many different ways.

DOBBS: You're the first president to talk openly and honestly about offshoring of production, the outsourcing of jobs, millions of jobs.

TRUMP: Terrible.

DOBBS: That has begun under your administration to come back to the country. The idea that made in America, by an American, hire in America, how—how likely is it that that can actually be realized with a business community that has fought on the wrong side of those issues for very long time?

TRUMP: Well, you're right, and it's surprising to me. For instance, NAFTA is horrible, one of the worst trade deals, it's just the worst.

. . .

DOBBS: When you focus first on free trade, people [say] "Oh, my gosh, he's talking about isolation"—

TRUMP: Unbelievable.

DOBBS: —and then they started looking at the numbers to see that we are—among all of our trading partners, that this country is the only one who said, please take what you want. We don't have a mature intelligent relationship with you. We don't have to have balanced trade. Every other major partner does precisely that.

TRUMP: The WTO, the World Trade Organization, was set up for the benefit of everybody but us.

DOBBS: Right.

TRUMP: They have taken advantage of this country like you wouldn't believe, and I say to my people you tell them, like, as an example, we lose all the lawsuits, almost all of the lawsuits in the WTO, within the WTO, because we have fewer judges than other countries. It's a setup you can't win. In other words, the panels are set up so that we don't have majorities. It was set up for the benefit of taking advantage of the United States.

The next question I asked the president was about a matter that would haunt his presidency for the next three years. Fresh news reports said that the Hillary Clinton campaign had created the Russiagate hoax by paying $6 million to a discredited British spy, Christopher Steele, who had produced a bogus dossier of dirt on the president. The FBI used it to win eavesdropping orders from the Foreign Intelligence Surveillance Court. Later, it would emerge that the FBI had erred and/or lied repeatedly to the court. I asked:

DOBBS: So how does President Trump feel about the bombshell revelations that the Democratic National Committee and the Democratic nominee's campaign are the ones who funded the smear campaign against presidential candidate Donald Trump, the man who would go on to win the presidency?

TRUMP: Don't forget, Hillary Clinton totally denied this and [said] she didn't know anything, she knew nothing. All of a sudden they found out. What I was amazed at is, it's almost six million [dollars] that they paid, and it's totally discredited. It's a phony call. I call it Fake News. It's disgraceful.

DOBBS: We'll be right back with a man who almost singlehandedly is changing the direction of this country, and changing it for the better. We continue live from Washington, DC, stay with us.

Cut to a commercial. Then the interview resumed:

DOBBS: The role of business in this country critical and fundamental to the country's well-being and its future to the creation of jobs. But business has taken such a voice in this town, in this swamp that you are the only countervailing influence to that dominance of US multinationals in this country. And the country owes you a great debt on so much but on that, in particular.

TRUMP: That's very interesting. It's well put. It's true.

The president closed by noting that "the good thing about social media" was that he could bypass the Fake News and reach 128 million followers on his platforms: "At least I can out—it's not that I want to do that, I'd rather not do it, I would love to not do it at all—but at least I can put out the truth, and I can put out the real word. And people agree."

Moments later we were ready to wrap.

DOBBS: I have to ask you, in conclusion, and I really appreciate your time, as does our audience—you came into this job fighting like hell, and you are fighting like hell every day.

TRUMP: More than I thought. I said I'm going to fight like hell. I didn't realize. It's even more, more so than I thought. But that's the way it is, and that's good.

DOBBS: And I've got to ask you, I mean, you are one of the most, I would say by the Left, particularly reviled, even hated man to ever hold your post.

TRUMP: I would say so.

DOBBS: You're also one of the most loved and respected.

TRUMP: I would say that, also.

DOBBS: In history. How does that feel? Where is Donald Trump today as you are just now beginning your presidency?

TRUMP: So the one thing that I really thought—because I thought it was treated very unfairly by the press during the campaign. With the win I said, the good news is now they'll start treating me well. But they got much worse. Lou, they put on stories on CNN and on MSNBC and CBS and ABC . . . which is ridiculous. They put on stories that are so false. . . . It is so dishonest, it is so fake.

. . .

DOBBS: Well, you are, if I may say, everything as advertised as you ran for president, and [we] appreciate everything you're doing. Thanks so much.

TRUMP: We're getting it done. I will promise you. We're getting it done, and this happened, a lot has already happened. Thank you very much.

The fundamental reforms President Trump won in his first three years in office will redound to our benefit for decades after he vacates the Oval Office. Some of his greatest hits: slashing thousands of burdensome federal regulations on business to unlock better growth; unshackling energy giants to make the United States the world's number one producer of oil and natural gas; prodding our partners in NATO to pay their required share of their own defense; pressuring drug makers to lower prices and stop soaking patients.

Other changes in the Trump era are less visible in reviving the United States and altering the arc of history in our favor. The sweeping overhaul President Trump has made in the federal judiciary, with the help of the Republican-run Senate, will gestate for decades.

Working from a list of eleven respected candidates that he released even before being elected, President Trump named two justices, Neil Gorsuch and Brett Kavanaugh, to the nine-member Supreme Court. He has appointed a passel of conservative federal judges: 44 appeals court judges (25 percent of the circuit court bench), and 112 district court judges (17 percent).

These are lifetime appointments. The youngest new judge, a woman, was just thirty-seven years old in early 2020; she could be issuing appeals court rulings for the next forty years.

Just as important, President Trump gave us back our swagger. We forget how bowed and bloodied we were in the opening years of the new century and beyond: the Tech Wreck and dot-com bubble burst of 2000; the 9/11 terror attacks in 2001 and wars in Afghanistan and Iraq that would end up costing the United States a total $2.4 trillion and almost 7,000 military personnel lives, killing almost half a million people in all.

Those conflicts would last twenty years. President Trump set a plan for pulling all remaining troops out of Iraq and came under blistering criticism from the liberal Left and Fake News, which usually applaud troop withdrawals. If he did it, it must be wrong.

Then came the Great Meltdown of 2008, the ensuing Great Recession, the $787 billion bailout of the "too big to fail" banks, and the $80 billion bailouts of GM and Chrysler—and President Obama's "apology tour" after taking office, traveling overseas to make speeches and give the bow, taking the United States to task for its past supposed sins.

The United States emerged as a nation with diminished prospects. The twentieth century was the American Century. The twenty-first century, it was quite clear at the time, was destined to become the Chinese Century. The conventional wisdom and media narrative told us we would just have to grit our teeth and put up with it. Eat it.

A "new normal" was at hand, it was declared. The forecast from the Federal Reserve and most everywhere else was bleak. Slack growth, tepid job creation, and a shift of economic power away from the United States and toward China were said to be inevitable. The same went for the further loss of manufacturing jobs to Asia, an underlying jobless rate that couldn't possibly fall below 5 percent, and a rapidly aging population and low birth rates. Forecasts said that artificial intelligence and machine learning could wipe out up to half of all US jobs in the next twenty-five years. Nothing to be done about it.

That is a sad recitation of mediocrity, "can't do," and "make do with less." It is an accurate portrayal of the *zeitgeist* at the time—the experts really believed those things. And when you settle for expecting less, you risk demanding less of our government and business leaders and pushing for less: be happy with crumbs; you get what you get, and you don't get upset.

Maybe that was the intention of those who espoused that outlook in the first place.

All that pessimism and cynical hopelessness, and Donald Trump arrived and washed it away with an overdose of enthusiasm and hyperbole. He changed the discourse from "It can't be done" to "Why not?"

He racked up an extraordinary string of wins on a dizzying array of fronts. Domestically: the biggest tax cuts in history, GDP growth 50 percent faster than what experts said was doable, record lows in the jobless rate, and a promising revival in US manufacturing. Real growth in wages after inflation rose for the first time in years. A jawboning feud pressured the Federal Reserve into cutting interest rates. And President Trump pushed back against Silicon Valley and the platforms that censor and suppress conservative voices and pursue a radical liberal slant.

There were international wins as well: new tariffs on imports of steel to force the Chinese to negotiate a new trade deal; tough new restrictions on China's unfair trade practices; a Trump treaty, the United States–Mexico–Canada Agreement (USMCA), that replaced the job-sucking North American Free Trade Agreement (NAFTA); devastating blows dealt to ISIS and Iran; controls on legal immigration that were upheld by the Supreme Court; the beginning of construction of

a new border wall; a crackdown on illegal immigration; and the opening of a new dialogue with the rogue regime of Kim Jong-un in North
Korea.

We won't even count as a victory—though clearly it is—the president's surviving multiple Russiagate investigations, being spied on by
the FBI, and the impeachment trial. It was an ordeal he never should
have faced, thrust on him by some of the most immoral, unethical politicians and Deep State bureaucrats ever to serve the American people.
Many of them should go to jail: Congressman Adam Schiff, House
Speaker Nancy Pelosi, Senator Chuck Schumer, FBI director James
Comey and senior agents at the FBI, CIA director John Brennan, former director of national intelligence James Clapper.

Ahhh, the thrill of victory. No wonder this will come to be called
the Trump Century. This is only a partial roster of the many times
President Trump was able to overcome the odds and win, helping his
country while sticking it in the faces of all of his naysayers—and often
gloating about it on Twitter in the process. Who can blame him?

Those wins resulted from something more than random, repeated
strokes of luck. Understand what it means to be so lucky as to be right
on so many issues, consecutively and constantly. Someone please do
the math for this; the chances must be one in a million. Face it, admit
it: this man was damn good at his job.

So many times his enemies and opponents said it couldn't be done,
President Trump pulled it off. Since taking office and embarking on
the Trump Century, he has been proven right just about every time he
clashed on big issues with the left-wing media and the ultra-Left. It's
the damnedest thing. It delights his fans and discourages his detractors. Think of how many times we have seen this:

• Manufacturing in the United States had been left for dead. Everyone said those jobs would *never* return. His predecessors had let it happen without comment or outrage. President Trump is good at outrage. He criticized any company that tried to shut down factories here, ridiculing them on Twitter. The pressure prodded foreign rivals into building here, too.

• In President Trump's first three years in office, the United States added almost half a million manufacturing jobs, almost double the 234,000 created during the last three years of the Obama administration.

• Critics said that the Trump tax cuts would benefit only the superwealthy. In fact, they left more money in the pockets of 65 percent of all Americans. The same goes for 82 percent of people earning $50,000 to $75,000 a year, according to the Tax Policy Center. Yet only 40 percent of us felt we had gotten a tax cut because of Democratic criticisms repeated by the innumerate media.

• Conventional wisdom said the US economy would limp along at 2 percent growth. In the second Obama term, we had nine quarters of sub–2 percent growth. That includes a low of just 1.26 percent growth in the second quarter of 2013. Yet GDP growth rocketed up to almost 3 percent in the first year of the Trump era. That equates to a $20 trillion economy growing almost twice as fast only a year later. It equates to an extra $280 billion in commerce in a year.

• Fake News and the experts predicted disaster when President Trump imposed tariffs on foreign imports to force China to the trade table. How did that work out? GDP kept growing, unemployment fell to 3.5 percent, and inflation stayed below 2 percent. China later agreed to buy an extra $200 billion worth of US-made goods.

- The climate change crazies flipped out when the president walked away from the Paris Agreement. It would have hobbled the United States with emissions caps while letting China churn out 30 percent of the world's greenhouse gases. Yet in the United States in 2019, emissions fell to the low levels of forty years ago. Utilities switched to cleaner and cheaper natural gas.
- His enemies insisted that he didn't have the constitutional authority to impose a ninety-day ban on travel from half a dozen Muslim-majority nations. The case went to the Supreme Court, which ruled: oh, yes, he does.
- President Trump authorized the mission in October 2019 that led to the killing of the leader of ISIS, the insanely evil terror group that Obama had once dismissed as "the junior varsity." Three months later, a US drone attack killed Iran's terror chief. Pundits predicted a wave of terror attacks and the risk of World War III. The predictions faded; there was no revenge, no war.
- When President Trump, however unpresidentially, jeered on Twitter at the dictator of North Korea, calling him "Rocket Man" and referring to the stubby man's short stature, experts warned of military confrontation. Didn't happen. Instead, our president struck up an odd peacemaking relationship with the reclusive despot, in part by courting him on Twitter.
- The border wall that candidate Trump promised is under construction, with more than 160 miles erected. Mexico hasn't paid for the wall, as candidate Trump fancifully promised. But our president did enlist Mexico's new president to station 15,000 Mexican troops on the Mexican side of the US-Mexico border to help stem illegal crossings.
- On October 13, 2019, Turkish troops were about to cross the Syrian border to attack Kurdish rebels. President Trump withdrew a thousand troops from harm's way, tweeting: "Very

smart not to be involved in the intense fighting along the Turkish Border, for a change. . . . Others may want to come in and fight for one side or the other. Let them!" The *New York Times* said it "could allow a resurgence of the Islamic State." CNN quoted a "US official" predicting a "second life" for ISIS. Didn't happen.

No one could have racked up so many right calls, so many times in a row, by making it up as they went along. Critics believe that the president has lurched from crisis to crisis, going by gut because he lacked an overarching strategy and the smarts to pursue it. They assumed that he operated without a belief system. The opposite is true.

Other politicians derive their talking points from polling and focus groups and million-dollar consultants. President Trump spent a few decades espousing many of the same views he holds today, honing his beliefs as a hotel/casino developer and later as the star of his own top-rated reality TV show, *The Apprentice*.

His central ethos of America first drives two constant goals: stoking growth and strengthening national security. Those three elements form the foundation of Trump strategy in five major areas, summed up in ten simple words:

1. Economy/jobs
2. Immigration/borders
3. Trade/China
4. National defense
5. Foreign policy

These five priorities interlock with one another strategically. In the president's view, to create more jobs (1), you must control immigration and your borders (2) while fixing bad deals in foreign trade, especially with China (3). These three things rely on a resolute and muscular military (4) and a strong foreign policy (5) to protect our interests and project our agenda around the world.

This five-pronged outlook explains the pursuits and policies of this president, why he does what he does, and why it is part of a premeditated, calculated, and well-thought-out strategy aimed at making the United States stronger, better, and more prosperous for all of us.

The president's many enemies missed this because, from the start, they had reduced him to one thing: a racist. They uniformly rejected his proposals, reverting to their echo chamber of moralizing, condemnation, and recalcitrance. How could any high-minded Dem ever sit down and negotiate with a racist? It may be even worse than that: many of the president's attackers were hurling charges of racism when they didn't believe it, at all; it was just an effective method of assault.

Their sanctimony pleased millennials and got lots of airtime on MSNBC and CNN—and it was what defeated them and hurt the credibility of the left-wing corporate media. Their obsession with fighting Donald Trump became all-consuming. Everything revolved around opposing him rather than pursuing their own bold new ideas.

What keeps President Trump moving forward? Ego and greed, the haters will say. He is driven by growth and ambition, business fans might argue. I see something deeper at work inside the man, something more spiritual, even if it might sound a little corny to the cynical souls among us. The essence of it is this: Donald Trump is an American patriot. Period. It is what motivates him. It drives him. It is his shelter in every storm, his horizon for every quest. He loves this country and everything it stands for, and he wants the United States to fulfill its destiny as the greatest nation on Earth. It offends him that we have been played for chumps by our trading "partners," by our allies in Europe, and especially by the biggest offender of all, China, and the frauds it has perpetrated on the United States for a very long time.

From that flows everything President Trump does. He feels lucky to have been born in the United States of America, lucky to have been brought up in wealth, and obligated to give back. He sees buttressing our national security and building a strong economy as the best way possible to lift up as many people as possible. I concur.

Thanks to the audacity and optimism of Donald Trump, the twenty-first century no longer belongs to China. It belongs to us. Even before the viral meltdown, China's economy was slowing and it was grappling, incompetently, with protests in Hong Kong. Then China gave the world the Wuhan virus, whether by nature or research lab. Before the coronavirus crisis, China was expected to eclipse the United States as the world's largest economy by 2030.

President Trump no doubt intends that the United States will hold on to its role as the largest, most resilient economy for a lot longer than that. In a world filled with globalist elites and apologists for China, he was the only leader to stand up and say "Enough!" In his ominous and darker view of our biggest rival, the Chinese were untrustworthy partners and irresponsible world citizens, prone to lying at every turn.

After Wuhan, the rest of the world would see that President Trump was right. Again.

CHAPTER TWO

RECOVERY OR RECRIMINATION

The Wuhan virus crisis was one of the worst cataclysms to strike the United States in its 250 years of history. It was also the biggest threat ever to the reelection prospects for a US president.

In early 2020, President Trump and the nation's governors called for the almost complete shutdown of the largest and most innovative and resilient economy on the planet. They did so based on the advice of the world's best medical experts and forecasts warning that 2 million Americans would die if we failed to act.

The nation complied. The result has been economic devastation.

The Wuhan pandemic—and our response to it—was all but certain to send the US economy staggering into a Wuhan recession. Only the severity and duration were in debate. This posed the most difficult challenge to face President Trump: getting reelected amid one of the deepest recessions we had ever endured.

It had been 120 years since a president had been reelected after a recession had taken hold in his election year: William McKinley, a pro-business Republican who had defeated William Jennings Bryan to win a second term in 1900.

President McKinley was assassinated nine months into his second term, at the Pan-American Exposition at the Temple of Music in Buffalo, New York. His vice president, Theodore Roosevelt, succeeded him. A day earlier, McKinley had made a speech urging reciprocity treaties that would open overseas markets to US companies. The problems President Trump (and I myself) have decried for three decades date back more than a century.

How ironic, even tragic, that an "invisible enemy," as President Trump called it, might deny him reelection and remove him from office, something the many efforts of so many passionately vicious adversaries had failed to achieve in the previous three years.

I had hoped the virus crisis might unite us behind our president just for once in his first term, even if in fear and with our newly shared sense of imminent mortality. The man had earned that from us— regardless of our political affiliation or any other kind of affiliation, for that matter. Unite us in a fight for our lives and the American way.

It was not to be. As the Wuhan pandemic engulfed our country, it was difficult to say who was the most outraged and vociferous Trump basher: the Democrats, the Fake News mob, or the Chinese government. The answer: it was a tie.

From the onset of the virus crisis, the Dems criticized President Trump's response and hatched plans to "investigate" it and second-guess his every decision. The henchman they picked to lead the partisan charade was Adam Schiff, the deceitful, bug-eyed California congressman who had spearheaded the Trump impeachment attack.

The left-wing corporate media were coconspirators in the effort. They obsessively cast doubt on the Trump administration's efforts rather than providing us with accurate, helpful information. Then China, seeking to distract from its incompetent and opaque response, began attacking the United States and President Trump for criticizing it. The Fake News reported it without so much as an arched eyebrow.

If at first you don't succeed, try, try again, as an old teacher's manual phrased it in 1840. The Never Trumpers and Fake News had tried the Hillary-funded spy dossier, Russiagate, Trump tax returns, the emoluments clause, and a plot to invoke the Twenty-fifth Amendment of the Constitution to replace him on the grounds of being emotionally and mentally unstable.

President Trump also overcame Obama-appointed judges who overruled him and were overturned on appeal, as well as a de facto coup attempt led by the FBI, CIA, and NSA, all of which authorized spying on his aides as a way to spy on him. It had the abiding approval of President Obama and Joe Biden as they were leaving office.

Trump survived, as well, the Robert Mueller investigation and its 448-page report, which led to zero indictments and a finding of "no evidence" of any American having colluded with the Russians. In fact, later it would emerge that there was a lack of *any* evidence proving the Deep State's assertions that the Russians had hacked into the computers of the Democratic National Committee. An insider DNC employee might have staged the hack. That would have destroyed the narrative crafted by the Deep State, the Dems, and Fake News for most of the Trump era.

He emerged victorious, too, from the Democratic impeachment trial over a single phone call to the new leader of Ukraine. That came from a supposed whistle-blower and Obama administration holdover in the National Security Agency who allied himself with Democratic coconspirators. Though his name was revealed and repeated across social media a million times, the Fake News lib media and Facebook et al. did all they could to block the public from learning his identity.

Now, in the aftermath of what many people call an act of God, the presidential antagonists—disrespectful, un-American verbal assassins—may finally have found a way to achieve the ouster of President Trump. Instead of stepping back and letting events occur naturally, they began deploying a new political tactic: recrimination. That impeded our recovery. The sanctimonious social justice warriors went against us and did all they could to steer us off course at a time when we needed all hands to pick up an oar and row.

The Wuhan pandemic put at risk our fundamental rights, beliefs, and hopes as citizens of the greatest nation on Earth. How we proceeded in the next few years would determine all that followed, and whoever we elected as the next president would have an indelible impact on that start.

My team at Fox was among the first to jump on the Wuhan story. We hammered China's culpability, impunity, and cover-up. We called it out for refusing the offer of President Trump to send in our best virus experts to help out. At one point, alarmed by the lack of testing kits in the United States, we confronted Trump's secretary of health and human services on my show. It is unclear whether he ever recovered. Joke.

The left-wing corporate media harassed the president, while my show prosecuted the real villain in the Wuhan emergency: China. It had delayed alerting the world, suppressed information, and stifled research. It continued to rebuff help from our own Centers for Disease Control and Prevention.

China was waging a cover-up that was dangerous and deceptive, as the prominent and respected China critic Michael Pillsbury told me on *Lou Dobbs Tonight*. On some nights, I told my viewers that China's violations were tantamount to an act of war. We warned that China's pandemic might have been an intentional first strike.

By contrast, Fake News fearmongers were putting their partisan slant on glaring display, live and unvarnished. At daily briefings, reporters lobbed verbal grenades at the president. Why hadn't he acted earlier? Does he feel guilty? How can he think he should be reelected? Why hadn't he gotten tested? Why wasn't he wearing a mask? How could he claim that the United States was leading the world in coronavirus testing? Why did he insist on referring to the "Wuhan virus" when doing so was racist?

The press pack should have been eliciting useful updates for the people. The Fake News fell short in reporting the progress made by the Trump administration's all-points response. Instead, it hounded the president and challenged every statement he tried to deliver to the American people. At the same time, it gave short shrift to the nefarious moves of China and its dereliction. It was as if the two were new allies.

Here is the story the lefty media was never willing to tell: President Trump's response to the virus crisis was epic in scale and boldness. It was decisive, inventive, and industrious. He signed bills for trillions of dollars in aid to businesses and US workers. Money started flowing weeks after the shutdown of the country. During the financial crisis of 2009, the new Obama team had taken nine months to pass a rescue package.

President Trump recruited industry to step up, confident that it could help solve our problems in better and faster ways. The administration lifted old regulations at the Food and Drug Administration to speed up the development of new vaccines, therapies, and testing. The president enlisted some of the biggest companies in the world, plying them with public praise, private pressure, and occasional "encouragement" on Twitter.

Dozens of corporations stepped up in a huge way. In a few cases, the president invoked the Defense Production Act to coerce manufacturers into switching their production to supply ventilators and related components. Ford and General Electric teamed up to produce 50,000 ventilators in a hundred days and then 30,000 a month. 3M pushed to double its production of N95 masks to 100 million units per month. Abbott Laboratories produced 50,000 virus tests per day. Also chipping in were the tech pillars Google, Amazon, Microsoft, and Salesforce.

That was only the beginning of the business response to the coronavirus. Thousands of companies helped in other ways by extending loan payments, reducing rent, setting up emergency response and information centers, offering special services for new stay-at-home users, and bombarding us with hundreds of ads assuring us that they were there for us. Which was okay at first, in smaller doses.

More important, President Trump provided encouragement and comfort to the American people in a time of terrible crisis. He assured us that things would return to normal and we would come back stronger. Franklin D. Roosevelt did it in fireside chats on the radio in the Great Depression. Donald Trump did it by tweet and in live briefings that yielded little chance for the Fake News libs to filter and block his message.

Americans needed encouragement rather than contentiousness. More than 3 million Americans lost their jobs in the first week of the virus crisis and the shutdown of the country; then almost 7 million more the next week and 6.6 million the week after that—more than 30 million people laid off in three months, with more to come, forced off the payrolls of businesses large and small in office towers, business parks, stadiums, concert halls, shopping malls, storefronts, restaurants, bars, health clubs, barbershops, nail salons, tattoo parlors, and more.

Even churches were ordered by the government to shut down, in violation of the First Amendment right to worship and freedom of assembly. Our loved ones were dying, alone and untouchable in quarantined isolation in the hospital, and then we were deprived of being able to gather together for a funeral to mourn their loss and celebrate their lives.

Estimates were that the coronavirus crisis would cause GDP to plummet by 30 percent in the first couple of quarters AC (after coronavirus). That is the equivalent of a $6 trillion decline in a year. Some Wall Street forecasters expected the damage to last until late 2021 or longer. Doomsayers think we may never come back.

That marked a drastic reversal in our collective outlook, which had gotten ever more upbeat and can-do in the Trump Century. After the election of Donald Trump in November 2016, the stock market began a historic rise that sent stock prices up an amazing 61 percent in thirty-nine months. The Dow Jones Industrial Average started at 18,332 and had soared to just 449 points shy of the 30,000 mark by February 12, 2020. That had created $11 trillion in new wealth in three years, which helped drive economic growth and lift our spirits.

After the United States came under attack from China and the Wuhan virus, stock prices tumbled almost 40 percent in six weeks. The Dow hit a shocking low of 18,591 on March 23, 2020. That wiped out *all* of the $11 trillion in stock price gains in a few weeks.

That had to hurt the president even more than the pussyhat protests, which had erupted after his election and had turned into an annual cry-fest. Remarkably, stock prices would bungee jump back up almost 30 percent by mid-May 2020 to approach the 24,000 mark. That was a sign of the markets' confidence in the Trump administration's response to the coronavirus. I liked the optimism, but then again, rebuilding America barely had begun.

As the crisis continued, cable news anchors, suitably somber and furrow-browed, served up constant updates of the body count as if this were Vietnam all over again. The nets put out reports of mass graves in New York, lines of refrigerator trucks for the spillover of bodies, and military ships put into place to assist overrun hospitals.

The media were less diligent in reporting that those stories often fizzled upon closer examination. They were #FakeNews, as the president would tweet, a term that he all but invented and that now even his opponents invoke. The grave sites had been in regular use, and they held zero coronavirus corpses. The military ships were treating a few dozen patients at the time, and by late April, the one docked in New York had departed. Happily, it had been little used. Some hospitals were all but empty.

Viewers also saw stories that were all too real: local cops wrestling a man off a bus for failing to wear an antiviral mask instead of just offering him one; other cops arresting a preacher for holding Sunday services, a lone surfer, and a father for playing T-ball with his daughter in a local park—for violating stay-at-home orders.

Some states let marijuana stores stay open and ordered churches to be closed. They forbade us to get a haircut in a small shop, yet let us stand in line at Costco. The Democratic governor of Michigan prohibited her constituents from growing their own food, banning the sale of seeds and gardening accessories as illegal and nonessential.

Yet few of us protested those unconstitutional overreaches initially. Millions of us cowered in fear, cloistered at home, hoarding hand sanitizer and toilet paper and barely going out at all. We accepted our lot. *What the hell was happening to us?*

That daily dose of turmoil may have had an upside both for the American people and for President Trump. Undecided voters had ample time during the lockdown to watch a lot of news live on cable and the broadcast nets and snippets on Twitter, Facebook, and YouTube. For millions of them, it may have been quite telling.

At press briefings daily at the White House, they could watch the president tangle combatively with Fake News. Each clash exposed more of their claws than his. His claws we already had seen many times. His getting the reporters to expose theirs so readily was more of a surprise. It may have played a role in the outcome of the 2020 election—especially if enough viewers were those in the forgotten middle, in industrial swing states such as Pennsylvania, Ohio, Wisconsin, and Michigan.

The Trump administration initiated its response to the coronavirus threat by forming a task force on January 28 and banning flights from China on January 31. Though the crisis loomed, the partisanship was unabated.

The travel halt from China brought a rebuke from Joe Biden, who was at a campaign event in Iowa on the same day. He told the crowd that Americans "need to have a president who[m] they can trust what he says about it, that he is going to act rationally about it. . . . This is no time for Donald Trump's record of hysteria and xenophobia—hysterical xenophobia—and fearmongering to lead the way instead of science."

It was an unintentionally revealing line. The Dems had displayed hysterical xenophobia and fearmongering in demonizing the Russians in their pernicious Russiagate hoax.

Two months later, in a full-blown crisis with the shutdown under way, Biden's deputy campaign manager, Kate Bedingfeld, told the media that Biden had supported the China travel ban: "Science supported this ban, therefore he did, too." That was patently false.

A day later, on April 4, the president told reporters at their daily joust that Biden had said he "was correct when I stopped people from China very early—very, very early—from coming into our country. . . . The other thing—so I appreciate the fact that he did, because I was called 'xenophobic,' 'racist.'"

The next day, Biden appeared on ABC News' *This Week* to gainsay the president's depiction: "You got to go faster than slower. And we started off awfully slow. He indicated that I complimented him on—on dealing with China. Well, you know, forty-five nations had already moved to keep—block China's personnel from being able to come to the United States before the president moved. . . . It's about the urgency. And I don't think there's been enough of it, urgency."

That was a deceptive and demented take on things. President Trump had been especially urgent in his response. Also, a total of thirty-six nations, rather than forty-five, had issued bans, and many were China's neighbors in Asia.

On January 22, 2020, I tweeted one of the first of our many notes on the Wuhan crisis:

> Lies Go Viral: @GordonGChang predicts there are more Coronavirus victims than China is admitting.

On January 30, I tweeted this praise for the president:

> Crushing Coronavirus: @DrMarkSiegel praises @POTUS administration for their swift and thorough action to protect U.S. citizens from Coronavirus.

On January 31:

> Controlling the crisis: @MikePillsbury says China is hurting themselves by denying @POTUS' gracious offer to help them combat Coronavirus.

On February 4, I cited the fiercest critic of China in the administration, Peter Navarro:

> Coronavirus's economic threat: Peter Navarro says the Chinese may use the coronavirus to dodge buying $200B in U.S. goods under the phase-one agreement.

On February 5:

Rejecting USA Help. @DrMarcSiegel says China continues to rebuff assistance from CDC members ready to travel to China.

On our program the evening of February 18, 2020, I said that I was genuinely impressed with the Trump administration's well-orchestrated response to the pandemic, telling Deputy Secretary of Homeland Security Ken Cuccinelli, "To the administration, and to the president for what you are doing because, within this country, it seems to be under control and every safeguard taken." Later in the show, I seconded my own emotion: "The government has done an amazing job, I think, in constraining the spread of the virus into this country." And indeed it had.

Still, everyone stumbles at times. Despite my clear admiration of and affection for this president, I am not aligned with anyone. I am a loner, and I don't like getting locked in with anyone of any particular stripe.

Two weeks after my praise for the Trump team, the nation was running behind in producing test kits so we could figure out how widespread was the virus. It felt to me as though a Trump cabinet member, Secretary of Health and Human Services Alex Azar, was being less than transparent and insufficiently urgent in his response. HHS had stopped short of upgrading Wuhan to a full-blown global pandemic.

When I questioned the HHS secretary about it on my show the night of Monday, March 2, 2020, he told me, "I'd say the message remains transparent, but one of confidence." I asked him, "Why not call it a pandemic, then, which you know very well it is. . . . Why not just be straightforward?" He argued that doing so would add nothing. He was about to give me the view of the World Health

Organization—the puppet of the People's Republic of China—when I interrupted him:

DOBBS: I'm not interested in their thinking, to be candid with you. You're responsible for your department. The NIH, the Centers for Disease Control are responsible for the public's health. . . .

I could care less what the World Health Organization has to say about what is happening to Americans and how this government is being led. . . .

Do we have tests that are now available, here in this country to the degree that we need them? Because the ratio, as I see it, there's a better ratio of testing for the coronavirus in other countries than we maintain here. We're screening few people. We're actually screening fewer people here because we don't have appropriate testing.

Secretary Azar responded that the CDC now had a validated test, and I pointed out that some tests were unreliable. He conceded that there had been some manufacturing issues in one stage of the test and insisted that the issue had been solved. Things then got a little testy:

DOBBS: Mr. Secretary, I don't want to play games with you, and I'm sure you don't want with me, but when you use words like "validated" and then say that one segment of their use was invalidated by problems, you know, it just—this is *not* transparency. . . .

AZAR: Well, Lou, actually, Lou, you don't want to hear it, but we've been transparent at every step of the way.

He assured me on air that massive supplies of tests were in production, now that an Obama-era rule had been lifted. I pressed again:

DOBBS: Mr. Secretary, I apologize. I apologize, but, frankly, you've just said that you had plenty, and now you're acknowledging that you didn't.

AZAR: No, I didn't.

DOBBS: I don't want to get bogged down in this. I mean, I really don't, but I *do* want to understand how bad is this going to get.

The secretary told me that the administration was planning for "all scenarios" and assured me that it would "take all measures needed to protect the American people." My concern was about transparency, as I told him: "I have absolute faith in the leadership. My concern is how much the American people are being told that they need to know, and that is, I think, everything."

The United States would soon produce millions of testing kits, prompting President Trump to brag about the numbers and send the Fake Media fact-checkers in frantic pursuit to prove that he had lied about it. On March 18, delighted by how well things were faring, we staged a poll of sorts on my program that was intentionally skewed: "How would you grade President Trump's leadership in the nation's fight against the Wuhan Chinese virus? Superb? Great? Or very good? Cast your vote on Twitter. We'd like to hear from you on this important question." If it were appropriate to add a smiley wink emoji in a book of this sort, I would insert one here.

The very same night, we also ran a searing segment on the CDC's getting billions of dollars in government funding with a mission statement of working 24/7 to protect Americans from domestic and foreign threats to health, safety, and security. As I told our audience:

DOBBS: The CDC, you might be curious to know, also uses tax-payer dollars on political issues that go well beyond fighting disease and in some cases support liberal ideology. Among some of those efforts, the CDC has spent money . . . to study gun violence, hosted a safe sex event with a porn star, held a transgender beauty contest, and built a $106 million visitor center filled with waterfalls and Japanese gardens. Very relaxing, President Trump. And we want to say thank you to David Horowitz for providing that list.

We give, we take away. Our support for the president's coronavirus response was a rare voice in the media. The corporate left-wing media, aka Fake News, devoted themselves to nipping at the heels of the president every time they could. Millions of voters saw that daily on live TV, as the president's interrogators fired one salvo after another. They sat six feet apart and in fewer numbers to comply with social distancing restrictions. The thinner ranks of antagonists pleased the president.

Like a good firing scene on *The Apprentice*, many of those real-life, on-air clashes may have worked to the president's benefit. His fans on Fox News averaged 2.5 million viewers for each briefing—and another 2.7 million people were watching on the two cable lib twins, MSNBC and CNN. That enabled him to reinforce his plaint about Fake News to millions of people, Dems and Republicans and independents.

The testy exchanges in the daily briefings on two back-to-back days, April 22 and 23, were illustrative. On the first day, the president admonished the *Washington Post* for pasting a sensationalist headline on a story in which CDC director Robert Redfield had said the flu season in the fall could complicate the coronavirus problem. Trump ushered Dr. Redfield to the lectern to address the topic for reporters.

The *Post* headline in question: "CDC Director Warns Second Wave of Coronavirus Is Likely to Be Even More Devastating."

REPORTER: And isn't that correct? Because—

THE PRESIDENT: That's not what he says.

VICE PRESIDENT MIKE PENCE: It's not what he said.

THE PRESIDENT: It's not what he said.

REPORTER: But if you have the two things happening—

THE PRESIDENT: The headline doesn't correspond to the story.

. . .

The testing problem. We've done more than any other nation in the world. Go a step further: If you added up the testing of every nation in the world, put them together, we've done substantially more than that. You people aren't satisfied.

So let's say we had 350 million people in the United States, right? Let's say. And if we gave every one of those people a test 10 times—so we give 350 people a test 10 times—the fake news media would say, "Where's the 11th time? He didn't do his job. Trump didn't do his job." Because you have a lot of bad reporting out there. It's very sad. And it's so bad—

REPORTER: But that's not true. That's not true. That wouldn't be the case—

THE PRESIDENT: But you're one of—you're one of the leaders of the bad reporting. You know?

REPORTER: No, but that's not true. I mean, this is—

THE PRESIDENT: Okay, let's get onto another subject. I wanted that to be—

REPORTER: Mr. President, can I just follow up on this real quick?

THE PRESIDENT: I wanted that to be cleared up. [Speaks about ramped-up ventilator production and how the media had failed to note or praise it.] . . . Nobody thought this could be done. The fake news was very unhappy that it was done. But you guys don't ask me about ventilators anymore.

REPORTER: Well, who's unhappy—who's unhappy that ventilators are being made, Mr. President?

THE PRESIDENT: Everybody [in the media]. Everybody. Because you never mention it. You never mention it. There's no story that's what a great job we've done with ventilators.

We're now supplying ventilators all over the world. Because no other country could have done what we did. And you should say that's a great story. Instead you say, "Trump was slow" or—slow? We were so fast.

Plus, we put the ban on so much earlier. When Nancy Pelosi, as an example—you don't say this—when she's having her rally in San Francisco—in Chinatown, in San Francisco. Nobody wants to say that. If we didn't—and Dr. Fauci said this—if we didn't close our country to China, we would have been so infected, like nobody's ever seen.

The president told reporters that he wanted to let Vice President Mike Pence speak now. First, though, he went on a 3,500-word Trumpian run of twenty points in a row, uninterrupted by anyone from Fake News. When I read the transcript of his remarks, it struck me as impossible that this ever could have been pulled off by Joe Biden as cable anchors interviewed him via Skype and a laptop while the nation was in quarantine

President Trump cited the revved-up production of ventilators, his early ban on flights from China, encouraging news that some states were reopening, a reminder about washing our hands, production of new tests, working with governors. Also: that he loves "those people that use all of those things," who go to tattoo parlors and other outlets, but "they can wait a little longer" before resuming business; the patriotic air shows in cities slated by the US Navy's Blue Angel Jets and the Air Force Thunderbirds—"they're incredible"; the coming July Fourth celebration on the Washington Mall.

And: the scale and speediness of the new Paycheck Protection Program and his successful prodding of Harvard and Stanford universities to return million of dollars in coronavirus assistance that they had applied for and received. Plus $30 billion in small-business loans to minority communities; $1.4 billion for 13,000 community health care centers for testing and treatment in underserved areas; $25 billion to expand testing, now at forty test sites, with dozens more coming online; and $7 billion for developing treatments, diagnostics, and therapies. Then the president wrapped up, all prompter, and tossed the ball to Mike Pence:

THE PRESIDENT: With love for our nation and loyalty for our fellow citizens, we will safeguard our families, care for our neighbors, heal the sick, protect our workers, and build a future for a country that is the greatest country anywhere in the world. And we're only going to get greater.

Thank you very much. Mike Pence, please.

One day later, unbowed by the president's blistering critique, the Fake News purveyors were at it again. New research had just shown that sunlight and heat kill the coronavirus with surprising lethality, which a government official presented at the press conference.

On surfaces, at 75 degrees Fahrenheit and in low humidity, the virus can live eighteen hours. It dies in only *two minutes* outdoors, at 75 degrees and 80 percent humidity in summer light. Likewise, using Lysol or another disinfectant aerosol, without summer light and at 75 degrees and 20 percent humidity, took an hour to kill the virus. Add summer light, and the virus died in only ninety seconds.

That was credible research conducted by the Department of Homeland Security's Science and Technology Directorate. The left-wing media all but ignored it and derided the president for having the temerity to hail the new findings. After the head of the directorate, William Bryan, presented the data, President Trump began playing spitball with advisors and reporters and pondering the possibility of new weapons against the virus:

PRESIDENT TRUMP: Thank you very much. So I asked Bill a question that probably some of you are thinking of, if you're totally into that world, which I find to be very interesting. So, supposing we hit the body with a tremendous—whether it's ultraviolet or just very powerful light—and I think you said that that hasn't been checked, but you're going to test it. And then I said, supposing you brought the light inside the body, which you can do either through the skin or in some other way, and I think you said you're going to test that, too. It sounds interesting.

ACTING UNDER SECRETARY BRYAN: We'll get to the right folks who could.

THE PRESIDENT: Right. And then I see the disinfectant, where it knocks it out in a minute. One minute. And is there a way we can do something like that, by injection inside or almost a cleaning. Because you see it [coronavirus] gets in the lungs and it does a tremendous number on the lungs. So it would be interesting to check that. So, that, you're going to have to use medical doctors with. But it sounds—it sounds interesting to me.

So we'll see. But the whole concept of the light, the way it kills it in one minute, that's—that's pretty powerful.

That was more than the Fake News could bear. The president had suggested studying those things in light of the new research. The Fake News story had the president advising Americans to inject Lysol to fight the Wuhan virus. He never said that.

To the White House press pack, that mattered little. The maker of Lysol, Reckitt Benckiser, put out a statement later that day warning that "under no circumstance" should its products be used internally to combat the coronavirus. The UK company said it was responding to "recent speculation and social media activity." Fake News blamed the president. From *The Hill*: "Lysol Maker Issues Warning Against Injections of Disinfectant After Trump Comments."

Moments later, the *Washington Post*'s Philip Rucker upbraided the president on national television for what he had just said. That arrogant nattering nabob of negativism (to quote Vice President Spiro

Agnew in 1970) presumed that he, rather than President Trump, rep-resented the American people. Here is the exchange:

RUCKER: Yes, Mr. President, after the presentation we just saw about the heat and the humidity, is it dangerous for you to make people think they would be safe by going outside in the heat, considering that so many people are dying in Florida, consider-ing that this virus has had an outbreak in Singapore, places that are hot and—

THE PRESIDENT: Yeah, here we go.

RUCKER: —are humid?

THE PRESIDENT: Here we go. The new—the new headline is: "Trump Asks People to go Outside. That's Dangerous." Here we go. Same old group. You ready? I hope people enjoy the sun. And if it has an impact, that's great. I'm just hearing this—not really for the first time. I mean, there's been a rumor that—you know, a very nice rumor—that you go outside in the sun, or you have heat and it does have an effect on other viruses.

But now we get it from one of the great laboratories of the world. I have to say, it covers a lot more territory than just this. This is—this is probably an easy thing, relatively speaking, for you.

I would like you to speak to the medical doctors to see if there's any way that you can apply light and heat to cure. You know—but if you could. And maybe you can, maybe you can't. Again, I say, maybe you can, maybe you can't. I'm not a doctor. But I'm like a person that has a good you know what.

RUCKER: But, sir, you're the President.

The president conferred with an advisor, then turned back to the reporter: "I think it's a great thing to look at. I mean, you know. Okay?" The reporter then invoked an old journo trick: say "respectfully" when you are being disrespectful:

RUCKER: But respectfully, sir, you're the President. And people tuning into these briefings, they want to get information and guidance and want to know what to do.

THE PRESIDENT: Hey—hey, Phil.

RUCKER: They're not looking for a rumor.

THE PRESIDENT: Hey, Phil. I'm the President, and you're Fake News. And you know what I'll say to you? I'll say it very nicely. I know you well.

ANOTHER REPORTER: Why do you say that?

THE PRESIDENT: I know you well.

Because I know the guy; I see what he writes. He's a total faker.

ANOTHER REPORTER: He's a good reporter.

THE PRESIDENT: So, are you ready? Are you ready? Are you ready? It's just a suggestion from a brilliant lab by a very, very smart, perhaps brilliant, man. He's talking about sun. He's talking about heat. And you see the numbers. So that's it; that's all I have. I'm just here to present talent. I'm here to present ideas, because we want ideas to get rid of this thing. And if heat is good and if sunlight is good, that's a great thing as far as I'm concerned.

Go ahead.

President Trump put up with that kind of abuse from the press almost daily. In all my years in the news business, never before had I seen the mainstream media gang up and so ardently antagonize the president of the United States. Never had I seen the media, always somewhat liberal, become so personal and disrespectful toward our president. It was as if Trump had broken them down and exposed their deepest flaws.

President Trump and his supporters would be on their own as he prodded Americans to return to normal life, navigating around the hindrances imposed by the Dems, Fake News, and China. He proceeded just as he had for his entire presidency. He was out to persuade the people, politicians, business leaders, organizations, institutions, universities, schools, and other members of society to join in the resumption of normalcy. His presidency and his legacy depended on it.

His defeat in the next election would be a severe setback for our country and our recovery. The president's haters, most certainly, were banking on it.

My prayer was that, by the time you read this, the Wuhan crisis will have passed us by, the casualties will have been a fraction of the darkest fears, and our recovery will be well under way. For the resistance, the confounding thing was that President Trump might triumph yet again. Even the Wuhan virus crisis ended up playing to his strengths.

We must rebuild America, and President Trump is a builder by nature. The pandemic revealed that a lot of what he had been saying for many years was on target: we must secure our borders and exercise tighter control over who enters our country. America first. No one else on Earth will look out for Americans as well as we ourselves will. He was right; multitudes of his opponents were wrong.

President Trump had shaped his America First agenda in forty years of logging unlikely wins and overcoming obstacles in business. At that point, his acumen had been sharpened by almost four intense years in the White House. The Wuhan crisis had reinforced his long-held beliefs and proven that he was right about a lot of things. Especially this: China is the real enemy; we cannot trust it.

CHAPTER THREE

CHINA LIED

China lied. People died.

That rhyme took hold on Twitter as the Wuhan crisis spread around the globe in 2020. China lied about the Wuhan virus and waited months to alert the world. When it notified the World Health Organization, it claimed that the virus posed no risk of human transmission. Then it lied about how many people had died in the contagion.

China also waged a disinformation campaign to blame the US military for possibly bringing the virus into the country. It diverted scrutiny by rallying the left-wing media to insist that the phrase "Wuhan virus" was racist against the Chinese.

When fear was exploding in the United States and President Trump bluntly criticized China publicly for its lagging response to the pandemic threat, Chinese officials threatened to withhold supplies of antibiotics and fever-reducing ibuprofen. That exposed China's endemic malevolent intent and its disregard for the rest of the world.

The shocker for many of us was that China provides 95 percent of our entire supply of those medicines. This must end. And never should we forget it. US drug makers made us dependent on China by switching their supply lines to that country to save a few bucks in labor costs. They did so without regard for the risk of relying on our archrival for mission-critical medical supplies. US manufacturers in all sorts of industries made that same unwise call, and the Wuhan virus crisis, if it has any upside at all, may shock many of them into bringing their production facilities back home to the United States. It's about time.

Threatening our access to medical supplies, lying about the Wuhan virus, blaming it on the US military, playing word games; these are the calculated, intentionally hostile acts of an evil empire rather than the missteps of an overly aggressive trading rival.

Admit it: President Trump was right all along. China is the real enemy. No matter what its government claims to be, China is the biggest threat to peace, prosperity, and world order. It is an irresponsible bully on the world stage. For decades it has been a crooked trading partner, a tyrant in business and international affairs, and a siphon of US jobs. More recently, it has been a rising military threat.

For thirty years, President Trump has been right about the unfair trade practices of our rivals and their devastating consequences for the US economy. Japan was his first target before its long slide into torpidity and stasis. Then he turned his sights on China as it grew ever more powerful. He has said he sees the two countries as interchangeable.

The larger and more powerful a country grows, the more fearful its rise can be for its neighbors and the rest of the world. When a nation is one of the biggest, most potent players on the planet, smaller rivals are all but helpless against it. The only thing they can fall back on is trust: Can they trust this leviathan to tell the truth, keep its word, and play by the rules? With regard to China, the answer is a clear no, and its leaders have shown that continuously for the last twenty or thirty years.

As China has arisen to become a global economic juggernaut, we tend to forget that it hews to the same habits and strategies the rulers of the Middle Kingdom have used for decades, even centuries. Those who know China intimately know that its government officials lie with impunity. This is official policy. When a crisis or controversy descends, the Chinese leaders obfuscate. They lie their asses off.

The Wuhan virus crisis has exposed China's penchant for political and economic transgressions, unfolding live and in real time as it reacts to daily events beyond its fiendishly dictatorial control. Its unabashed, ballsy proclivity for fabrication is on display, bared more than ever before. China may be the most formidable opponent to confront any US president in our history.

Donald Trump, in turn, is the most formidable president ever to confront China—and the *only* president to do so. He has been waiting for this face-off for much of his life, long before he ever became president. Their intertwined futures were always on a collision course. From the start of his presidency, he raced to bring the Chinese to heel while fending off his enemies at home and elsewhere. After Wuhan, his confrontation with China took on new stakes as we debated over how widely and how rapidly to open up the economy for recovery.

The contest would play out on two different levels: in the real world of restoring the people's confidence, rebuilding the economy, and shaping the logistics of crisis response; and in the online world of Twitter, the pixel punditry, and hordes of his followers and resistors. The online world is where President Trump excels—and, it turns out, it is where China, too, would devote a lot of effort.

In the aftermath of accusations and second-guessing, President Trump got squeezed between his Chinese rivals and Never Trumpers as they found common ground. During the crisis, China would launch a sophisticated social media campaign with woke, umbrage-laden messaging that would appeal to the left-wing corporate media and the woke, liberal social justice warriors leading the Trump resistance in the United States.

China needed new friends. It was reeling from the shocking economic and reputational repercussions of its failure to thwart the most frightening new virus in a century. Soon it began seeking allies, whether anyone knew it or not. The rest of the world had every reason to view China with hostility and suspicion in the aftermath of the Wuhan catastrophe. So China decided to play the victim, and many of the events that would follow were an outgrowth of that posturing.

As huge an economy as China was able to build, as powerful as its military buildup was, suddenly the virus crisis turned that massive, egotistical nation-state into an oversensitive adolescent, seething with resentment and taking offense where none was intended. In fact, the more powerful China gets, the more thin-skinned its leaders become about its reputation—and the more they will lie to protect it from harm.

Long before the Wuhan pandemic rocked the United States, my nightly program on Fox Business, *Lou Dobbs Tonight*, was one of the harshest and bluntest critics of China's many sins. After Wuhan, we turned the criticism up to high heat.

The Chinese government richly deserved it. Usually when a nation screws the pooch on a disaster of global scale so terribly, the right thing to do is to apologize abjectly and focus on the fix. Instead China grew obstinate and accusatory as never before.

That had its intended effect: a lot of meek souls in the media retreated from challenging China aggressively, which was more reason for us at my show to take the reverse approach. On January 22, my Twitter account put out one of the first of our many notes on the Wuhan crisis, saying "Lies Go Viral" and predicting that China was underreporting cases.

Eight days later we praised the president:

Crushing Coronavirus: @DrMarkSiegel praises @POTUS administration for their swift and thorough action to protect U.S. citizens from Coronavirus.

Mike Pillsbury came to our attention on January 31, just as President Trump was announcing a ban on all incoming flights from China, by saying that China was hurting itself by denying Trump's gracious offer to help it combat the coronavirus. A few days later Peter Navarro was on our show saying that China might use the coronavirus as an excuse to dodge buying $200 billion in US goods, a deal signed only weeks before.

In February, things got ever more aggressive in our hard look at the Wuhan crisis. Expert guests including my Fox News colleague Dr. Marc Siegel and Peter Navarro helped amp up the pressure on China. We were outing the outbreak, as we put it, and we warned that 80 percent of the active elements in US drugs were produced in China, that the country's infection and death numbers couldn't be believed, and that the Wuhan crisis was being worsened by that nation's failing leadership.

On February 16, I posted a tweet linking to a new story in the *Daily Mail* in London and asked anyone who was watching: Did coronavirus originate in a Chinese government laboratory?

A few weeks later, the China machine shot back. On March 4, a tweet appeared in the Twitter feed of a New York–based China scholar at the Council on Foreign Relations, Yanzhong Huang, Sr. It included a web link to an op-ed just posted on the website of the Xinhua news service, the government's primary propaganda outlet: "China's Xinhua News just posted a piece titled 'Be bold: the world owes China a thank you,' which says if China imposes restrictions on pharmaceutical exports, US will be 'plunged into the mighty sea of coronavirus.'" There was no other comment.

Click on the link, and a Chinese-language version of the article shows up. Translated into English, the op-ed pointed out that China makes ingredients that go into 90 percent of the US pharmaceutical supply, as well as most of the medical masks used in the United States. It said that China could retaliate by banning exports of both. If it did so, the United States would "fall into the sea of new coronaviruses" and "fall into the hell of a new coronavirus pneumonia epidemic." The Trump flight ban, it said, was "unscrupulous."

It was an astonishing set of threats by the government that had just spawned a biblical-scale plague upon the planet. China should have taken responsibility, accepted the penalty, repented, and asked for our help. Instead, it offered condemnation instead of contrition, arrogance in the place of humility.

That was an unmitigated outrage. It struck me as an act of war. The US media worked far harder at second-guessing President Trump and trying to thwart much of what he was achieving on our behalf.

On April 16, 2020, on my nighttime program, I'd had enough. I interviewed a regular on my show, K. T. McFarland, who had been General Michael Flynn's deputy national security advisor in the early months of the Trump administration. I told her my view of China's actions:

> *If we don't go to war over the loss of 31,000 now, and certainly more to come, 31,000 American lives, what do we go to war over? When do we quit sending strong letters and talking tough? At what point are there consequences for this kind of behavior? Because whether they did it intentionally or not, we do know this: that that virus was unleashed on the world and they lied, and that is the same as making it an intentional and conscious act of warfare, as far as I'm concerned.*

My argument became truer still as the United States surpassed 100,000 deaths by mid-May 2020. President Trump was publicly saying that he would stop short of "retaliation" against the Chinese for the Wuhan pandemic. If the pandemic didn't warrant retaliation by the United States, however, then almost nothing does; perhaps the president didn't rule out "punishment" by sanctions and other measures. China must pay for this, in some way and in some form of recompense.

Four days later we called China a global pariah and warned that the Chinese Communist Party would face consequences for its lies. What else were the Chinese hiding? One of the foremost experts on China and its many misdeeds was Michael Pillsbury, a longtime associate of mine and a former Department of Defense official. On my show on April 21, he told me that the president's investigation of China's role in the crisis might lead to serious ramifications for WHO for its complicity in the Wuhan cover-up.

Another erudite and fearless China critic on my show was Gordon Chang, a New Jersey–born Chinese American who studied at Cornell Law School and lived in China for almost twenty years as a lawyer for two white-shoe firms. He argued that China had recklessly and maliciously ensured that the virus would spread to sicken people around the world.

He had a good point about that. China had disclosed the virus outbreak on New Year's Eve 2019. In the ensuing four months, more than *430,000* Chinese travelers arrived on direct flights from China to the United States. That included 40,000 people who were able to land in our country in the two months *after* President Trump imposed the ban on flights from China.

No wonder President Trump has been willing to fight through so much static to tighten the control over our borders and who is allowed to visit or immigrate here.

On April 30, at an event in the East Room, the president stepped up his accusations aimed at China. The gathering was on behalf of elderly

Americans, who were dying in disproportionately higher numbers in the pandemic. The president had previously theorized that China might have unleashed the virus in some kind of horrible mistake. Now he added that its release might have been intentional.

And that might be the case. I had underscored this on air a number of times in the crisis. But the president's mere speculation was reported as an overstep by the Associated Press, which shares news and op-eds with hundreds of member media outlets. The headline read, "Trump Speculates That China Released Virus in Lab 'Mistake.'"

The story implied that Trump advisors doubted his theory and might have felt pressured to produce findings to support it—the latter assertion coming from no one quoted in the story at all. The AP also quoted Chinese officials at length, printing their lies without any attempt to present them skeptically.

The AP story said, "This all comes as the pace of Trump's own original response continues to come under scrutiny, questioned as too meager and too slow." It invoked the Chinese statement that those claims were "unfounded and purely fabricated out of nothing" and that the lab "strictly implements bio-security procedures that would prevent the release of any pathogen." It quoted a researcher at Scripps Research in La Jolla, California, who put the odds of a lab accident at "a million to one." How could she know that?

The story neglected to reveal that China had refused all outside efforts to visit the lab and inspect it and had rejected offers of help by the CDC. The AP did say that China "criticized those in the US who say China should be held accountable for the global pandemic, saying they should spend their time on 'better controlling the epidemic situation at home.'" It was as if the AP were subtly chastising our president.

Also, the AP added that "a U.S. intelligence official *disputed the notion* that there was pressure on agencies to bolster a particular theory."

(Italics added.) Whose notion, ventured by whom and from where? And if that anonymous government spook had "disputed the notion," why did the AP put it into the story at all?

This kind of sleazy innuendo has riddled thousands of news articles and op-eds on dozens of lib media platforms. The renegades were raging out of control, and they were destroying the credibility of all the media. The Trump battle cry of "Fake News" came to be invoked by millions of Americans on the right *and* the left.

The virus's origins trace to Wuhan (population 11 million), five hundred miles west of Shanghai. In April 2020, the Chinese government was claiming that there had been only 2,700 deaths in that city. That is a small number, considering that New York City had more than 12,000 covid-19 deaths at the time, in a smaller population (8.5 million). Then, in one day, the Chinese government increased the death count in Wuhan *precisely* 50 percent, from 2,579 to 3,869. The real toll may be higher. Media reporters spotted many thousands more urns lined up at Wuhan's eight funeral homes.

The Chinese government also made use of "useful idiots" to distract the media from the country's own culpability. When a Chinese official tried to lay blame on the US military for bringing the virus into China, the mainstream media reported it with a straight face—without denials from the Trump administration and the US military.

China did that on the record—on the president's chosen platform of Twitter, no less. It was a clear in-your-face. The BBC has identified fifty-five Twitter accounts run by Chinese consulates, embassies, and diplomats, thirty-two of them opened in 2019 alone. In August 2019, Twitter disclosed a significant state-backed information operation originating from within China and targeting pro-democracy (and anti-China) protesters in Hong Kong. Twitter closed almost a thousand accounts and suspended 200,000 others as being illegitimate, *The Guardian* reported at the time.

That part of the Chinese propaganda campaign began on March 12, 2020, with a tweet from Zhao Lijian, a Chinese Foreign Ministry spokesman. The platform is blocked inside China, but the Chinese flack, as the *New York Times* put it, "has made good use of the platform . . . to push a newly aggressive, and hawkish, diplomatic strategy." The tweet plumbed conspiracy fears:

When did patient zero begin in US? How many people are infected? What are the names of the hospitals? It might be US army who brought the epidemic to Wuhan. Be transparent! Make public your data! US owe us an explanation!

That brazen, scandalous claim, made on the record by a Chinese government spokesman, reveals the unbridled arrogance that China's leaders now feel in their role on the world stage. They no longer bother trying to hide it. They pulled the stunt in the middle of a frightening emergency that was killing Americans by the thousands.

The tweet was reposted to the Chinese social site Weibo, where it was viewed 160 million times within a few hours. In the United States, the Chinese spokesman has 600,000 followers. The video attached to the Chinese tweet was viewed almost 4 million times. The tweet gave the impression that it summed up the video, when, in fact, they were unrelated.

The China-edited, one-minute snippet, set to haunting music added in postproduction and overlaid with screen captions in Chinese characters, shows CDC director Robert Redfield testifying to Congress. His comments have nothing to do with "patient zero," tracking case numbers, or the US military.

A congressman asked him whether, if someone had died of the flu a few months earlier without being tested, he or she might have died of the coronavirus. According to Redfield, "Some cases have actually been diagnosed that way in the United States today." From that, China speculated, "it might be US Army who brought the epidemic to Wuhan."

The post drew more than 15,000 "likes" on Twitter and almost 8,000 retweets by people (or software bots) forwarding the Chinese charges onward to their followers.

The left-wing Fake News media, including the *New York Times*, Reuters, CNN, and dozens of other lefty outlets in the United States, relayed that ridiculous canard to the American people. The *Times* published a story of 1,500 words and twenty-seven paragraphs, without quoting a US military spokesperson denying the assertion. Reuters ran a story of 600 words and eighteen paragraphs, without including a US denial.

The mainstream media largely avoided stating the obvious: that the China charge was silliness on its face. US troops would have had to take the long way around the barn. The virus had come from a breed of bats native to parts of China. The bats may have contaminated the food of a pangolin, an especially scaly Chinese armadillo. Edible meat from the pangolin may have been consumed by people at a "wet market."

Actually, that would be the long way around the barn, too: in Wuhan, the government runs two virus labs that are believed to have been working with the strain. What I want to know is: Why were they working with this wickedly lethal virus at all? Maybe it was research for germ warfare gone awry.

Further, the China tweet brimmmed with subconscious persuasion. It opened by asking "when" patient zero had showed up in the United States, rather than *whether* the disease had begun here at all, as if that were already accepted as true. The term "patient zero" made it sound very clinical and medical, and the exhortation to "Be transparent!" should have been directed at China itself.

In the United States, Twitter quashes conservative commentary and suspends conservative accounts for transgressions the libs get away with all the time. Yet a spokesman for China, the most antagonistic government on the planet, is welcome to spin away all he wants.

The Chinese government also concocted a controversy over what to call the virus. In late January 2020, it began playing the race card. Suddenly, invoking the word "China," "Chinese," or "Wuhan" and linking it to "virus" or "pneumonia" was racist. Anti-Asian.

That was inane—and entirely insane. The world pandemic was raging, and China had been the epicenter and botched the early response that might have avoided all of this. It was scrambling to prevent the deaths of possibly millions of people. Yet it took time out to start strong-arming the media in the United States, the United Kingdom, Australia, and other parts of the world to press its concern about what everyone was allowed to call it.

You will note in this book that I use the terms "Wuhan virus," "China virus," and the like many times. My point is that we should refuse to surrender to verbal brainwashing and that no change in terminology can hide the truth it is intended to hide: that the virus came from China.

The new word game was a cunning parody of the woke ultra-Left and Fake News in this country. Many viruses carry place names: Ebola (a river in the Democratic Republic of Congo), Spanish flu, Zika (a rain forest in Uganda), MERS (Middle East respiratory syndrome), Norovirus (Norwalk, Ohio). Plenty of diseases, too: Lyme disease (Connecticut), German measles, Coxsackievirus, Rocky Mountain spotted fever, and the Guinea worm (for a spot on the west coast of Africa, a phrase coined in the seventeenth century).

George Orwell said that if you control the language, you control the masses. Now the Chinese were doing him one better, covering up a crime on a global scale.

One early effort in the word-banning campaign showed up on January 29 in a Chinese-language editorial on the China-controlled Xinhua news site. It argued that the term "Wuhan pneumonia" was "prejudiced" and disrespectful to the city's residents. A week later, the *Global Times* weighed in similarly. The tabloid newspaper is an affiliate of the Chinese Communist Party's *People's Daily*. Recently the *Global Times* staff had conducted "astroturfing" campaigns against foes of the Chinese government, faking a grassroots movement and hammering such targets as the artist and antigovernment activist Ai Weiwei.

The *Daily Telegraph* in London reported that the *Global Times* was one of three state-controlled media outlets that were "flooding Facebook and Instagram with undisclosed political adverts whitewashing its role in the coronavirus pandemic and pinning blame on Donald Trump." The other two players were Xinhua and China Central Television, and the trio "have targeted users across the world with promoted stories in English, Chinese, and Arabic."

The ads "depicted Mr Trump as misguided and racist, and suggested that the virus might have originated in the US," the *Telegraph* reported. The Facebook and Instagram ads "initially ran without a political disclaimer, allowing them to hide information about who they were targeting and sometimes letting them sidestep Facebook's strict rules on political advertising."

On February 6, while that effort was under way, the *Global Times* published an op-ed by a New York writer. At one point, the writer referred to the philosopher Susan Sontag and her musings on "the need to make a dreaded disease foreign." He added, "Some Chinese people in the US and European countries have been verbally or physically attacked by hotheaded strangers simply because they were wearing masks." This sounds short on evidence. Some? How many?

Then the *Global Times* column cited a new Change.org petition that had drawn 70,000 signatures to demand apologies from the *Herald Sun* and *Daily Telegraph* in Australia. The first paper's sin was "for its reference to 'Chinese virus' in a subtitle under a headline." The second one for "highlighting the words 'China kids stay home' in a headline." That is kind of strict, yes?

For the kicker, the writer cited a story in a Chinese-language newspaper in New York that told of the "ordeal" of a couple rushing back to the United States from their wedding in China to beat the travel ban. "They said they now understand how people from Muslim countries on the US travel ban feel. These are small sparks carrying hope."

For China, anyway.

Coda: the writer of the op-ed is a New York–based reporter for *Sing Tao Daily*, a Hong Kong–based newspaper that supports the government of China.

By February 12, the Chinese government's campaign was in full hue and cry. Xinhua posted an "opinion" under the headline "Western Media Should Quit Racist Reporting as China Fights Epidemic." The writer complained that "certain western media outlets are generating viciously misleading reports, undercutting global efforts to end the epidemic."

Here is an example of what the Chinese Propaganda Ministry considers to be "racist reporting" so vicious that it undercuts the government's efforts to quell the Wuhan pandemic: *Der Spiegel* in Germany featured a cover image of "a man wearing a red hoody, protective masks, goggles and earphones, with a giant headline 'Coronavirus made in China.'" A newspaper in Denmark ran "a cartoon of the Chinese national flag that replaced the five symbolic stars with virus-like particles." Clever.

Last, from *People's Daily*: "Not to be outdone, the *Wall Street Journal* last week published an op-ed that called China and its people the real 'Sick Man of Asia,' a highly racist tag that has long been spurned."

Though a less hysterical view emerged in the *South China Morning Post* by a columnist who traced the origins of the phrase "Sick Man of Asia" to more than a hundred years ago to describe China as a crumbling empire like the original "Sick Man of Europe," the Ottoman Empire.

The *South China Morning Post* put an angry headline over such a nuanced piece: "Coronavirus Triggers an Ugly Rash of Racism as the Old Ideas of 'Yellow Peril' and 'Sick Man of Asia' return."

The paper, it should be noted, is owned by Alibaba, which is pro-government (and it is impossible to build an Alibaba in China and be anything but pro-government).

Any nuance was beside the point. China's feelings were hurt, and it wanted to play the victim and make a show of it. The government was so miffed by the headline in *Wall Street Journal* that it expelled three of the paper's Beijing-based reporters in response, the biggest such move it had made since the Mao era, the paper reported. That was way off target. The headline ran with an opinion article that had nothing to do with the news reporters China had just jettisoned.

The *Journal* reported, "China's Foreign Ministry said the move Wednesday was punishment for a recent opinion piece published by the Journal."

It is a rather histrionic overreaction to an innocuous slight that the Chinese leaders could have let pass. It wasn't as if the *Journal* had accused China of intentionally releasing the Wuhan virus from a state-run germ warfare lab in the city. Which . . . someone really ought to investigate.

The Chinese let nothing pass these days. Their newfound wealth, accumulated over twenty years of cheating on free-trade deals, has made them cockier and more combative. They might want to work harder on trying more velvet glove and less iron fist. In the coronavirus crisis, it has been all of the latter.

In April 2020, government officials in Australia were watching the alarming rise in virus cases and deaths. They did what they always do in a crisis: expressed grave concern, monitored the situation, set plans to investigate things on the ground. How dare they say investigate? China immediately threatened a quid pro quo: back off, or the largest nation in the world will stop buying your beef. The headline in the pro-government *Sing Tao Daily* was "Australian Ambassador Advocating Investigation of New Crown Epidemic Situation: May Not Buy Australian Beef."

The Chinese paper reported that the government's ambassador to Australia had given an interview to the *Australian Financial Review*, in which he fired a warning shot that threatened broad retaliation. An official transcript of the interview, posted in English by the Chinese Embassy in Australia, quotes the Chinese ambassador to Australia as saying:

"I think if the mood is going from bad to worse, people would think why should we go to such a country while it's not so friendly to China. The tourists may have second thoughts. Maybe the parents of the students would also think whether this place, which they find is not so friendly, even hostile, is the best place to send their kids to. So, it's up to the public, the people to decide. And also, maybe the ordinary people will think why should they drink Australian wine or eat Australian beef. Why couldn't we do it differently?"

Beef, wine, tourism, foreign exchange education, the Chinese official had just pulled off threats to four separate industries in Australia in a single sentence. It was like a great tumbling run in an Olympics floor routine. China has stopped even trying to conceal the threats and coercion that it carries into every negotiation. It is doubtful, without explicit orders from the Chinese government, that the Chinese "general public" would ever be so upset by the issue as to forgo a nice steak and a glass of cabernet.

It's hard to remember that before Trump, the ruling class believed that capitalism would eventually bring the Chinese government around to being a good global citizen. How could it ever have imagined such a thing would come to pass?

China is "giving the sell," as the practice is called in fake pro wrestling. When your opponent flips you onto the canvas, you let out an anguished scream and writhe in phony pain to "sell" the stunt as the real thing. That was what China was doing during the coronavirus crisis, manufacturing phony umbrage over supposed "microaggressions."

It is a bold strategy. The rest of the world should be furious with China for inadvertently (or intentionally) giving coronavirus to the world. So China cries racism and turns itself into the wailing victim. It is as if China's leaders had been studying the tactics of America's radical Left. Everything is a racist slight.

Nobody seemed to mind that this was entirely contrived as a propaganda ploy by the Communist government in China. The Fake News gleefully advanced the China memes. As the crisis grew more frightening with each day, reporters at televised briefings challenged President Trump directly on why he insisted on using such hurtful terms. There had been "dozens" of recent attacks on Asian Americans in a nation of more than 300 million people. The president, as ever, obstinately held his ground, as in this back-and-forth at a White House briefing on March 18:

CECILIA VEGA OF ABC NEWS: Okay. Why do you keep calling this the "Chinese virus"? There are reports of dozens of incidents of bias against Chinese Americans in this country. Your own aide, Secretary Azar, says he does not use this term. He says, "Ethnicity does not cause the virus." Why do you keep using this? A lot of—

THE PRESIDENT: Because it comes from China.

VEGA: —people say it's racist.

THE PRESIDENT: It's not racist at all. No, not at all. It comes from China, that's why. It comes from China. I want to be accurate.

VEGA: And no concerns about Chinese Americans—

THE PRESIDENT: Yeah, please, John.

VEGA: —in this country?

THE PRESIDENT: No.

VEGA: And to the aides behind you, are you comfortable with this term?

THE PRESIDENT: No, I have a great—I have great love for all of the people from our country. But, as you know, China tried to say at one point—maybe they've stopped now—that it was caused by American soldiers. That can't happen. It's not going to happen—not as long as I'm president. It comes from China.

Five days later, the case numbers and deaths were beginning to build, but the press pack was still in hot pursuit of the president for continuing to use the W-word. At the briefing on March 23, another reporter circled back around, this time asking the president whether his saying "Chinese virus" was fueling a rise in assaults on Asians.

REPORTER: Mr. President, just quickly, a second question: What prompted you to say at the beginning of your—beginning of your comments that you're going to take care of the Asian Americans? Has there been something in particular that was prompting you?

THE PRESIDENT: Yeah, because it seems that there could be a little bit of nasty language toward the Asian Americans in our country, and I don't like that at all. These are incredible people. They love our country, and I'm not going to let it happen. So I just wanted to make that point—

REPORTER: Do you think you contribute to the (inaudible)?

THE PRESIDENT: —because they're blaming China. People are blaming China—

REPORTER: —by calling this the "Chinese virus"?

PRESIDENT: —and they are making statements to great American citizens that happen to be of Asian heritage. And I'm not going to let that happen.

Once the Fake News plays the same old race card, it opens the way for more attacks. At one point at a Trump coronavirus briefing, a black reporter for NPR directed a question to Surgeon General Jerome Adams: Was he aware that some people felt that his use of the terms "big momma" and "pop-pop" for "grandmother" and "grandfather" was condescending and racist? The surgeon general himself is black.

It is doubtful that the lib lemmings in the left-wing corporate media would ever have come to the angle of racist-naming the virus if the Chinese government had not fabricated a nonsensical and unnecessary controversy around it. They were happy to prosecute it, though, for the government of our avowed enemy.

Many of the same liberal media outlets dogging the president on his supposedly racist terminology were using certain terms themselves before China cowed them into a new level of political correctness:

WASHINGTON POST, JANUARY 8, 2020: "China Virus: Specter of New Illness Emerging from Wuhan"

NEW YORK TIMES, JANUARY 15: "Japan and Thailand Confirm New Cases of Chinese Coronavirus"

NEW YORK TIMES, JANUARY 21: "First Patient with Wuhan Coronavirus Is Identified in the U.S."

WIRED, JANUARY 22: "Experts Can't Agree If the Wuhan Virus Is a Global Crisis"

THE GLOBE AND MAIL, JANUARY 23: "Wuhan Virus Death Toll Rises to 26 as China Moves to Restrict Travel in More Cities"

CNN, JANUARY 23: "Wuhan Coronavirus Is Not Yet a Public Health Emergency of International Concern, WHO Says"

WASHINGTON POST, JANUARY 31: "An expert on influenza says we may not be able to contain the Wuhan coronavirus—but it may not spread as fast as some others."

CNN.COM, FEBRUARY 8: "A US national in China is believed to be the first foreigner to die from the Wuhan coronavirus, authorities confirmed."

Even the same Chinese government–controlled websites that decried the use of "Wuhan virus" and other terms as racist were using those same words right up until Chinese officials decided to declare them to be racist and xenophobic. And hurtful, of course, very hurtful. Some examples:

XINHUA, JANUARY 9: "Wuhan's Viral Pneumonia Epidemic with Unknown Causes"

PEOPLE'S DAILY, JANUARY 21: "Fourth Death Confirmed in Wuhan Pneumonia Cases"

GLOBAL TIMES, JANUARY 22: "Wuhan Pneumonia a Wake-up Call for Basic Chinese Research"

PEOPLE'S DAILY, JANUARY 25: "Wuhan Pneumonia a Wake-up Call for Basic Chinese Research"

By early February 2020, however, the Chinese media websites had scrubbed all their online content to remove the supposedly offending words. You won't find any trace of them if you look up those old stories on their sites today. In their eyes, it never happened.

Were all these left-wing media outlets racist for using that term— or did they let the Chinese rewrite the rules in the middle of a crisis? And all because the Chinese government wanted to divert attention away from its own monumental failings. The truth underlying both sides—the lib media wretches *and* China—was that they shared a common enemy in President Trump.

He had barely gotten started. The president's advisors were planning their next steps and bracing for further confrontations with the dissembling liars of the Chinese government. In the previous two years, President Trump had figured out the Chinese mind-set and had racked up a string of remarkable victories against our largest rival. It was another challenge that no one had thought he could win. Up next, an inside look.

CHAPTER FOUR

A SILENT FIRST STEP

For thirty years as a businessman, Donald Trump hewed to the steadfast belief that the United States was under unfair economic attack from its partners and rivals in global trade. Imposing a tough new regime of tariffs was always the centerpiece of any strategy that might emerge.

When he ran for office, he vowed again and again to slap tariffs on China. That defied the orthodoxy of the Washington establishment and the rest of the world, which had hailed "free" trade as the panacea for ensuring economic growth and lifting up what we once called the Third World—before that term, too, was banned by the word police and deemed politically incorrect.

That kind of one-world condescension has cost us hugely, and the Chinese continue to exploit it. President Trump has been haranguing us about it since the 1980s. To the former New York real estate deal-maker, the most humiliating thing of all is getting suckered in a bad deal, and China had been suckering us from the get-go.

China, and Japan before it, had used a muscular industrial policy to build their bases in manufacturing and technology, picking industries they hoped to dominate one day and deploying a range of subsidies, surtaxes, incentives, penalties, and trade barriers to give their companies an advantage over foreign rivals. When they were small and plucky, that was viewed as allowable: let them rise.

The problem was that they and other nations continued their industrial policy practices even after they had grown up and were thriving. Meanwhile, the United States eschewed establishing any kind of assertive industrial policy of its own. President Trump wanted to reverse that.

Thus, from the moment he moved into the White House, even before the Wuhan crisis overtook the nation and his agenda, he was determined to confront China's egregious excesses. He already had a plan in mind. It would rile his critics and grant him extraordinary, freewheeling tariff powers as he grappled with China's government. Shrewdly, he would take that first step to impose new tariffs and change world trade by resorting to a law passed fifty-five years earlier that included a provision so rarely invoked that it was considered a dead letter.

President Trump's declared intent to slap tariffs on China flew in the face of the orthodoxy preached by the high priests of free trade: big business and the Chamber of Commerce; big-labor unions (which represented less than 7 percent of private workers); the heads of the Republican and Democratic parties who were in the room agreeing to the fiasco; economists at the best universities in the world and multinational corporations; and the New Oligarchs of Silicon Valley, who were keen on importing cheap engineers from overseas.

The battle went beyond an arcane policy agreement. It was President Trump's shock-and-awe attack on the globalist elites who had fattened themselves on free trade while American workers had suffered the consequences. He owed them nothing special; they had done nothing to elect him. That made him even more of a mortal threat to them.

The president was braced for the pillars of the establishment to go all out in opposing him. In that regard they exceeded his expectations. Three days after he moved into the White House, he set off the first of many kerfuffles to come by summarily scrapping the plans of his predecessor and both parties to join eleven Asian nations in the Trans-Pacific Partnership (TPP).

A chorus of critics cried out. They jumped past debating policy to question the new president's intelligence, experience, and wisdom. The opposition took the move personally, sulking at the offense. How dare President Trump do something in the opposite way from the way everyone else before him had done it? How dare he simply toss aside, on his third day in office without even thinking about it, the diligent work of so many hundreds of trade experts and congressional staff? They actually said that.

One of the nastiest critics was the late senator John McCain of Arizona, the revered Republican gray head and Vietnam War hero. A couple of weeks later he called President Trump's plan "stupid" and berated other Republicans for failing to stand up and blast away in similar fashion.

Moreover, he did so premeditatedly, in a formal statement quoted at length on the *Business Insider* website: "I'm more than a little outraged. . . . As Trump has gone on to all kind of other stupid things in the trade-policy arena, the relative silence on this issue from the congressional leadership is appalling." He complained that the United States' role as a leader of an open trade system, "something that generations of American leaders have worked to build and protect ever since the Franklin D. Roosevelt administration," was being "trashed by the president of the United States and the congressional leadership is silent. What the hell are they thinking? Or are they just pusillanimous cowards?"

Business Insider reported that Joshua Meltzer, a wonk at the Brookings Institution, a center-left think tank, had declared, "This is the first administration that utterly misunderstands trade. I mean they've just got it basically and utterly wrong." He'd said that Peter Navarro, the new Trump advisor and fierce hawk on China, was "just wrong, flat wrong. I'm not the only person to say that—every economist on the right and left says the same thing. And so, Trump misunderstands trade, and his economic advisors do."

Another economist, Lee Branstetter at Carnegie Mellon University, who had helped hammer out TPP terms as a member of President Obama's Council of Economic Advisers, told *Business Insider*, "The fact that all that effort has come to nothing is really quite infuriating." He and the reporter were apparently unaware that his objection was hilarious.

For as long as anyone in trade could remember, the superior form of international deals had always been multilateral accords that brought together as many parties as possible to agree on terms across as wide a region and as many products and industries as possible. That had been seen as intelligent and efficient: one size fits all.

It was, however, the converse of what President Trump had in mind. The Trans-Pacific Partnership accord had been aimed at putting up a stronger united front against China by uniting the United States with other Asian countries, but even Democrats were cooling on it. The new occupant of the White House preferred one-on-one bilateral trade deals—and he had been waiting a long time to confront China in his own distinctive way.

Four months later, the Trump administration quietly took a first step toward imposing new tariffs on China. It was the start of building a transformative new trade policy for the Trump Century. The left-wing corporate media paid little attention.

On April 27, President Trump ordered Wilbur Ross, the distressed-assets buyout billionaire whom he had named secretary of commerce, to launch an investigation into whether imports of steel and aluminum posed a threat to national security.

It was a misdirection play. We are the number one *importer* of steel in the world, in part because we lack the natural resources to make it from scratch. We bring in 30 million metric tons each year—and none of it comes directly from China, the number one *exporter* of steel to the world. China has half the worldwide steel market, and it is the number one producer of aluminum, with two-thirds of the global output. Our allies Canada, Brazil, and South Korea are among our biggest providers. No matter; Commerce officials believed that China was hurting US steel producers. And it was.

Now, to take the first step, President Trump was invoking the Trade Expansion Act of 1962, which was aimed at empowering the president to *reduce* tariffs and negotiate free-trade agreements. A particular provision, Section 232, granted the president the power to hike tariffs on national security grounds. That provided him with the loophole he needed.

China had developed far more steel production capacity than its economy could possibly use, and it was dumping the excess supply onto the global market. That had depressed prices everywhere else in the world. Moreover, Chinese steel was arriving in the United States anyway, after first being transshipped to third-party countries to sidestep the penalties that the United States had imposed on China steel some years earlier for the same kind of trade violations.

The journey to a new regime of tariffs on China had begun. Few observers understood that at the time, not even the president's own advisors. Most of them opposed the idea of tariffs, especially so early in this new presidency. A second step toward confronting China came on May 17. The US International Trade Commission, the independent, bipartisan panel that advises the administration, launched an investigation of unfair trade practices in the US solar panel market.

The ITC was responding to a request filed by SolarWorld, a small maker of solar panels in Hillsboro, Oregon, and an Atlanta-based manufacturer named Suniva. Suniva had already gone bankrupt and fired all two hundred of its employees. Both had struggled with brutal price competition from China rivals. This time, the ITC investigation invoked a different provision of a different law: Section 201 of the Trade Act of 1974.

That was a hidden gift to President Trump. Typically, when a US company files a complaint against overseas rivals, the US government takes the case to the World Trade Organization. The WTO initiates an investigation that can drag on for years. It rules on the merit of the case, often going against US firms. Appeals slow the process.

President Trump has called the WTO biased and blatantly anti-American, with judges from other countries predisposed to ruling against us. I concur emphatically.

By contrast, the SolarWorld complaint, under Section 201, was a "safeguard case," and it granted the president wide power to help an industry in trouble. It hadn't been used since George W. Bush had invoked it in restricting steel imports in 2002, a year after the terrorist attacks on 9/11.

Now, by invoking Section 201, President Trump had the power to declare tariffs on Chinese solar imports solely in his own hands. That gave him an early chance to prove that tariffs were the best way to force our trading rivals to renegotiate terms that had disadvantaged us for years. For a president who had been calling for tough tariffs on trading bullies such as China for much of his life, it must have felt exhilarating.

The Great Ripoff of America started from the moment the United States, in the year 2000, voted to provide "permanent normal trade relations" to China, giving it "most favored nation" status. That committed us to granting China the same terms we gave even our best longtime allies. Yet China wouldn't have to do much to open its markets to us. US politicians and trade negotiators led the way in agreeing to that, spawning two decades of steady, chronic economic damage here at home.

It was a historic misjudgment on the part of our government leaders, scandalous for the decades of deleterious effects it would wreak on US workers and their families. As I have said many times in many venues, free trade has been the most expensive policy this nation ever pursued. There is nothing free about ever-larger trade deficits, mounting trade debts, and the loss of millions of good US jobs.

The vexing thing is less that they made that error and more that even after its draining consequences began to surface, the contented policy elites, big-business fat cats, and navel-gazing economists of the establishment continued to insist they were right. We told you so, gentlemen (for most of that time, few women were on hand to bear any of the blame).

President Trump likes to keep score. For him, the size of our trade deficit with China and the rest of the world is equivalent to the Dow Jones Industrial Average for the stock market. The trade deficit is the measure of how much more in goods we buy from China and other countries each year than they buy from us.

A trade deficit of zero, totally in balance, is ideal. A trade *surplus* would be even better; it would mean that other nations were buying more in US goods than they were selling to us. Their capital and their demand for more goods made in the United States would spur hiring and stronger growth in this country.

Instead, we have sent $10 trillion *more* overseas than our foreign trading partners have sent back to us in the past twenty years. That is how much more we spent on foreign-made products and services than overseas partners spent on products made in the United States. It has been a massive drain of cash, capital, energy, jobs, and innovation out of the United States and into the rest of the world, especially China. This has deprived us of capital and hurt the growth of our factories, assembly plants, offices, and stores.

Citizen Trump saw the soaring trade deficits as the clearest evidence that the United States was losing a trade war that had been under way for twenty years. Our leaders were chumps, bad negotiators who had been suckered by craftier leaders and other players who had parroted free-trade slogans and covertly advanced their own national interests. The trade deficit was a scorecard—and the United States was losing.

For consumers in the United States, the shortsighted upside was the low prices for goods created by China's flooding the American market with supercheap products, including the latest in electronics: HDTV screens the size of a ping-pong table for a pittance; the handheld supercomputers now known as smartphones. They often bore American brand names but were made by Chinese hands. This supposed benefit to consumers may have sedated the people while our policy leaders failed us—but never did it sedate me. I knew we were losing badly by the day in the one-sided exchange that diminished the United States and made *China* great.

President Trump's bent placed him in the center of a gladiator's coliseum, surrounded by a frightful array of people with the most to lose if we toppled the old order: giant corporations, yet also powerful labor unions; the Democrats, but also the venerable Republicans who had been complicit in creating the problem; liberal think tanks and conservative ones, too; and, as always, his unruly Fake News Greek chorus.

They went after President Trump even before he won the election.

As the Republican National Convention was nominating him in July 2016, the *New York Times* published an "analysis" of the Trump tariff plan. It described the conventional view:

> *Mr. Trump's framing is in direct conflict with the view even of many economists who are sympathetic to the idea that current trade arrangements work to the detriment of American workers and want to see change. In the more widespread view among economists, trade deficits are not inherently good or bad; they can be either, depending on circumstances. The trade deficit is not a scorecard.*

Sniff, sniff.

China is the biggest part of a larger problem for the United States. Even our allies have engaged in unfair trade terms that penalize the United States. This is why, simultaneously, the president also pursued that sore point in separate deals with Mexico, Canada, Japan, the European Union, and the United Kingdom. This man likes to negotiate one-on-one.

Currently, the US trade deficit with the entire world runs from $600 billion to $700 billion a year. China consumes roughly half of that massive flow of capital out of our country. Our trade deficit with the world as a whole—how much more money we are paying to other nations for their goods than they are paying us for the goods they buy from us—less than doubled in twenty years. But our deficit with China soared more than fivefold.

China now sells the United States a total of $450 billion in goods annually and growing, while the United States sells China only $100 billion a year. In the twenty-year period from 1999 to 2019, China's share of the total US trade deficit rose from 21 percent to 56 percent of the total trade gap.

Thus, China has benefited far more than any other country from the explosion in international commerce and unfair "free trade." And the United States has fared far worse than any other nation.

In 2019, that imbalance produced a US-China trade deficit totaling $345 billion. This means that every three years, we are sending China $1 trillion *more* than China sends back to us. We are richly funding the country that wants to supplant the United States as the leader of the world.

President Trump spent years hammering on that—as did I—in the style with which he approaches everything: bluntly, relentlessly, and repeatedly. He blames previous US administrations, rather than China and other countries, for this debilitating drain on our assets. Our leaders let this happen.

Along the way, most US business and political leaders expressed little concern. It mattered not a lick whether Democrats or Republicans controlled the White House, the Senate, or the House of Representatives; the slogans of so-called free trade dominated both political parties and the boardrooms of corporate America. The trade policies of Al Gore, George Bush, John McCain, Barack Obama, and Mitt Romney were all but indistinguishable.

The growing trade deficit was earnestly reported in newspapers such as the *Wall Street Journal* as if it were a weather report—something simply to talk about. Economists, even those who knew better, mostly kept quiet about it. They were high priests of the cult of free trade and wanted to avoid riling the American people.

As I have said before and will repeat, few honest economists practice the craft, and none of them works for a US corporation. The ones who do are propagandists hired to preach the free-trade gospel that will ensure their giant employers an ample supply of cheaper labor overseas. If that hurts American workers, too bad: shareholder returns, right?

Other nations around the world, including Germany, Japan, and China, impose serious limitations on US imports and protect their own national industries. By contrast, our government spent decades pursuing a policy of openness in trade, with near neutrality as to whether one country or region was reaping the upside at our expense.

Nobody called for an intervention to demand that our trading partners stop the cheating. The topic didn't come up on the daily talk show circuit or in the editorial pages of the major newspapers.

The enormous increase in the US trade deficit failed to grow into an issue even in the presidential debates that took place every four years. This slack attitude was captured perfectly at the party nominating conventions for the 2008 presidential campaign when Obama ran against McCain. The Republican Party platform on trade said, meekly and weakly, "we need to be at the table when trade rules are written to make sure that free trade is indeed a two-way street"—ignoring the fact that US leaders had been there at the start, approving the very policies that were killing us.

The Democratic platform on trade was laughable. Already, with Obama politics pushing the party even further left, fashionable lib concerns were overwhelming the traditional Democratic attachment to the American working class. The Dem platform plank sounds like the opening essay for a high school yearbook. Or a hippie commune:

We believe that trade should strengthen the American economy and create more American jobs, while also laying a foundation for democratic, equitable, and sustainable growth around the world. Trade has been a cornerstone of our growth and global development, but we will not be able to sustain this growth if it favors the few rather than the many. We must build on the wealth that open markets have created, and share its benefits more equitably.

In other words, the Democratic Party's position was that the United States owns too much of the world's wealth—never mind whether we *earned* it or not—and the government should endeavor to share it "more equitably" with the rest of the world. The Dems also wanted to ensure that the wealth of the richest people in the United States could be collected in greater amounts and redistributed to the rest of us. This stops short of policies that could help create new jobs so we can earn more money on our own rather than taking it from Jeff Bezos.

This has enriched China and starved the United States of domestic investment, making our companies dependent on foreign investment—and putting them under foreign control. Diverting production to cheaper labor markets overseas has shrunk US manufacturing and suppressed wage growth.

It has also contributed to layoffs and restructurings that left millions of people out of the workforce and no longer even bothering to look for a job. The collateral damage of lower wages and less job growth leads to more people applying for government assistance, making more of us more dependent on it.

Everyone in power allowed this to happen, in part because they had a hard time seeing it and in part because the worst impact showed up in the blue-collar "flyover" states that the people in power barely cared about. The most damaging effects came in towns in Ohio, Pennsylvania, and Michigan, while service jobs were growing in urban markets such as New York, Los Angeles, and Chicago.

This consequence of the trade deficit adds to the federal *budget* deficit, how much more money the government spends than it takes in each year. When the economy fails to grow adequately, the government has to borrow more to pay out more in stimulus packages. Our economy was so battered after the Great Recession of 2009 that it limped along for years instead of racing into recovery, as it had in the wake of prior economic crises.

The yawning trade gap with our largest rival would be a tough but fair outcome if US companies had had a fair shot at competing and had faltered. In the 1980s, when US automakers were near bankruptcy because of competition from better-made and better-priced cars made in Japan, they begged for protectionism and tariffs. Yet the legendary Lee Iacocca, a friend and mentor of President Trump, led Chrysler to a revival without them.

We lost the trade battle in the decades that followed, and unfair trade terms were only one factor. Another aspect was the apparently easy decision by US companies to cut and run rather than find a way to make things work back home. Past presidents bragged about their retraining programs for displaced workers while collecting campaign contributions from CEOs who were sending jobs overseas. Trump's predecessors did little to shame them for it or find solutions to reverse the corrosive trend.

They were too sophisticated and intelligent to do that, too well briefed by the experts who had designed the dysfunctional system in the first place. President Trump took a different route. Even before he took office, he dished out blistering criticism of Carrier Corporation, the HVAC giant, after which it scrapped plans to move 1,400 jobs to Mexico. Liberals pointed out that three years later, Carrier moved many of those jobs to Mexico, anyway. However, tell that to the people who had kept their jobs for three more years and the executives who had realized that sending jobs to Mexico would have been a business failure to be ashamed of. Who knows how many companies scrapped relocation plans to avoid the same PR debacle?

President Trump knows that tariffs can help on this front, too. The new regime was only the starting point rather than the end goal. President Trump was intent on tackling myriad other weaknesses and outrages in US trading terms with China, in particular.

Even when China granted US businesses access to its market, its terms were onerous. For the US imports it did let in (soybeans, air-craft, and microchips), it charged far higher tariffs than the low tar-iffs the United States applied to imports from China on smartphones, computers, and apparel.

When China did allow US firms to set up shop inside the country, it imposed a system of forced corporate marriages. They were required to form joint ventures with Chinese companies and hand over major-ity ownership to them—and share intellectual property and technol-ogy secrets with them. At times, Chinese firms would set up shadow factories and turn out knockoffs in direct competition with their US partners. US companies endured the abuse silently, hoping things would get better.

The world sat back and let China behave that way with impunity. Both Democratic and Republican administrations in the United States looked the other way and did nothing to call it out. US multination-als, always lustful for the great potential of the huge Chinese market, were silent for the most part. They complained to Washington only privately, if bitterly. Everybody was high on globalization.

The gnawing thing was that those same complainers took a long view and maintained that ultimately, the free-trade structure of the past would pay off in the end; China would mature on the global stage and amend its evil ways. Neither of those predictions came true. They were laughably naive.

Donald Trump had always argued that tariffs would force China to repatriate some of the trillions of dollars it was sucking out of the US economy; pay us back. As the world would soon observe, the Trump tariffs provided even more advantages.

The mere threat of tariffs would give a new incentive to other nations to get out in front of the trading crackdown and grant better terms to the United States. Some foreign firms would see more upside to building factories in the United States, rather than shipping their products in from overseas. Tariffs would make the latter option more costly.

The new policy also would put US companies on notice. They might then prefer to build in the United States rather than import from overseas. US giants knew by that time that if they dared try to shutter a US factory and move production to China or elsewhere, they risked a bitter rebuke from the president—on Twitter, no less, where ridicule and scorn spread with viral intensity.

That risk loomed large, even if moving production overseas or buying from foreign suppliers made the best economic sense from a profit-and-loss standpoint.

More important, because of the unusual tactic of invoking national security concerns and Sections 232 and 202, President Trump now had a new stack of bargaining chips at his disposal.

The loophole gave him wide wiggle room to impose new tariffs, as well as increase or decrease them. Or he could delay or extend them and decide which nations would be affected and how much. All of that without the interference of the World Trade Organization or the fractious gridlock of Congress.

That would enhance President Trump's position at the negotiating table and, of course, create supposed fears in the resistance that the president was overstepping the powers bestowed on him by the Constitution.

The first test lab for the effort was the US solar panel industry. By now it should have been thriving with a couple million jobs. In the Obama era, the nascent industry got a huge boost in the stimulus package passed in response to the financial meltdown in 2008.

Congress had authorized $787 billion in new spending to help the economy rebound. The sum was considered to be massive at the time, although it would be dwarfed a decade later by the cost of recovering from the Wuhan shutdown. Tucked inside the stim pack was an injection of $90 billion in easy government money for green energy, including $27 billion for the fledgling solar industry in the form of loan guarantees, grants, and tax incentives for green products.

It was intended to seed an entirely new US industry, one that would create millions of green jobs in solar and wind energy, electric car batteries, and more. Solar energy was such a naturally perfect idea that adherents had been on a quest for the right solutions since the 1970s. They had never gotten that far, though, because the price of oil, coal, and gas had never risen high enough to make green energy a competitive option.

Now the federal government's new largesse finally would deliver on the green revolution and create enough jobs to replace the almost 10 million jobs in the US fossil-fuel energy industry. Those evil carbon emitters.

The real beneficiary, as it would turn out, would be China.

No wonder that would be one of the first areas targeted for tariffs by the Trump administration. It was a twofer: a way to both land a direct hit on China and neutralize the fallout from a bloated Democratic program. Dozens of solar start-ups popped up in the United States to get in on the gold rush. The volume of solar energy production capacity installed annually in the United States more than tripled from 2012 to 2016. But the Americans got slaughtered.

In the same period, imports of solar panels grew by *500 percent*— fivefold, quintupled! Prices fell by 60 percent in that period, driven by artificially cheap imports from China, "to a point where most U.S. producers ceased domestic production, moved their facilities to other countries, or declared bankruptcy," as the ITC later reported.

Twenty-five companies making solar panels in the United States went out of business in five years. That left only two US producers of solar cells and modules—one of which exited the business in

2017—and eight firms that produced modules using cells imported from—yup—China.

By 2012, even before the big Obama push, China's share of the global solar cell market, at just 7 percent in 2005, had soared to 62 percent. It now controls almost 70 percent of the total global expansion capacity.

If Chinese manufacturers had simply outrun US companies, outcompeted them, that would be fine. But their win was a result of *unfair* trade, just as President Trump said it was. Industrial policy and unfair government subsidies had created China's competitive edge, the ITC later ruled. That violates the tenets of international trade.

Struggling solar firms began appealing for help to Commerce and the ITC almost a decade ago, but it was like playing a game of whack-a-mole. The International Trade Commission would assess tariffs on solar panels from China, and makers would shift their production to another country. That enabled them to avoid the tariffs and continue their predatory pricing. Target that new country, and they would pack up and move again.

In 2011, Commerce had ruled that China was subsidizing its producers to sell panels below cost in the United States. A year later, it imposed antidumping penalties and so-called countervailing duties, "but Chinese producers evaded the duties through loopholes and relocating production to Taiwan," an ITC report said.

So domestic makers filed new petitions to close the Taiwan loophole, and, as the ITC noted in 2013, "Chinese producers responded by moving production abroad, primarily to Malaysia, as well as Singapore, Germany, and Korea."

At that point, the investigation into whether steel imports should be tariffed was well along. That was battleground number one. The SolarWorld case was battleground number two. Then a third front opened in the Trump team's China strategy: two weeks after the ITC began investigating China in the solar panel market, Whirlpool filed a Section 206 "safeguard" complaint of its own with the ITC, on May 31, 2017. This time the target was white goods: washers, dryers, and other large household appliances.

The grounds Whirlpool cited aligned perfectly with the steel argument: that Whirlpool had been harmed by imports and South Korean companies were avoiding tariffs by shipping their washers and dryers from other countries. Once again, the safeguard complaint handed tariff power directly to President Trump. As in the SolarWorld case, the Whirlpool review would provide a model for going after China and levying new tariffs on steel.

Whirlpool was escalating a long-festering trade feud with the South Korean makers Samsung and LG Electronics over their supercheap prices in the US market. The two brands had consumed a 35 percent share of the market in the US, compared with Whirlpool's 35 percent share. Twice before, Whirlpool had brought trade cases against the two companies, charging that they were illegally dumping washing machines on the US market below their real cost. Each time, Whirlpool had won—but nothing had changed. Another game of whack-a-mole.

The first complaint to the International Trade Commission was filed in 2012. It targeted Samsung and LG washers made in Korea and Mexico, alleging that they were subsidized by the Korean government and dumped on the US market at artificially low prices. A year later, the ITC ruled in Whirlpool's favor and imposed countervailing duties on washing machines made in Korea and Mexico.

So Samsung and LG moved their production to China, thus evading the new tax. In 2015, imports of washers and dryers from China surged, and Whirlpool filed a second complaint. In early 2017, the first year of the Trump era, the Department of Commerce issued an anti-dumping order and ordered US Customs to begin collecting taxes at the border on washers arriving from China.

Next, Samsung and LG moved their production from China to Vietnam and Thailand. Frustrated and with a new sheriff in Washington, Whirlpool executives changed tactics. Instead of seeking action against imports from Vietnam and Thailand, Whirlpool filed a "safeguard petition" on May 31, 2017, asking the US government to impose global tariffs on *all* washer and dryer imports—from everywhere. Five days later, the ITC started its investigation.

The US company argued that the South Korean makers were avoiding the duties by shipping products via other countries, depressing prices and unfairly hurting Whirlpool back in the United States, just as China's cheap steel prices in international markets hurt prices here, despite a lack of shipments directly from China itself.

And then nothing. The three trade investigations—steel imports, solar panels, washers and dryers—plugged along invisibly during the second half of 2017. Meanwhile, the Trump team prepared for other efforts to topple the trade protocols of the past with even our friendliest allies: Japan, Mexico, Canada, and the European Union.

Some Trump advisors continued to hope he was just bluffing about the tariffs. On July 7, 2017, three months after the Section 232 case on steel began, the *New York Times* ran a story quoting an advisor to the Trump campaign, Stephen Moore, a Heritage Foundation economist, as saying "If he actually pulls the trigger, it could be highly disruptive to world trade. It's not even going to really work in terms of helping American workers."

As the president likes to point out, economists are usually wrong.

A month later, President Trump took a fourth significant step toward imposing the Trump tariffs. On August 14, he signed an executive memorandum instructing US trade representative Robert Lighthizer to look into launching an investigation into allegations that China was stealing US intellectual property.

A signing ceremony—for the rather mundane act of requesting an investigation—was set up at three that afternoon in the Diplomatic Reception Room of the White House. Even the Fake News media dutifully reported it, including the *New York Times* and the *Washington Post*. President Trump made the most of it. "The theft of intellectual property by foreign countries costs our nation millions of jobs and billions and billions of dollars each and every year," he declared as he scribbled his huge, crowded signature with a black Sharpie at the bottom of the executive memo. "For too long, this wealth has been drained from our country while Washington has done nothing. They have never done anything about it. But Washington will turn a blind eye no longer."

He added that "we will protect forgotten Americans who have been left behind by a global trade system that has failed to look—and I mean look—out for their interests. They have not been looking out at all."

On a roll now, he went for a big finish, all teleprompter. "We will safeguard the copyrights, patents, trademarks, trade secrets, and other intellectual property that is so vital to our security and to our prosperity. We will uphold our values, we will defend our workers, and we will protect the innovations, creations, and inventions that power our magnificent country."

It was a great moment for a president in his first year, and he had gone through a very rough few months. His efforts to abolish Obamacare had foundered on a defection of several US senators from his own party. Now he was taking a big step toward keeping a founding promise of his campaign and his presidency.

Though the signing ceremony in August 2017 elicited little

coverage of the actual policy move, the next action the president took, five months later, would unleash an avalanche of media histrionics and waves of new opposition and outrage.

On January 22, 2018, he announced tariffs on all international imports of solar panels and modules, washing machines, and dryers. In solar panels, he imposed a levy of 30 percent on the price of all imported solar cells and modules in the first year, declining by 5 percentage points in each of the next four years. In white goods, he slapped a 20 percent tariff on the first 1.2 million washers imported into the United States and a 50 percent tariff on everything over that, plus a 50 percent tariff on washer parts.

Whirlpool's CEO was thrilled, putting out a statement celebrating "a victory for workers and consumers alike." From most other quarters, though, oh, the agony. "A great loss for American consumers and workers," said Samsung.

The requisite cascade of concerned commenters intoned soberly about the adverse impacts the move would cause. China's minister of commerce called the Trump move "an abuse." Mexico, which exports more than a billion dollars' worth of Samsung washing machines and other brands to the United States per year, vowed to use "all legal resources available" to fight the new tariffs.

South Korea's trade minister said they were "excessive and a clear violation" of world trade rules; he vowed to file a complaint with the World Trade Organization. Two days after the Trump tariffs were unveiled, the country did so. It took the WTO sixteen months to rule on the case, and on May 6, 2019, the United States lifted the antidumping and countervailing duties on imports of washers from South Korea.

Special interests weighed in, too. Howard Crystal, an attorney for the Center for Biological Diversity, a green nonprofit based in Arizona, solemnly declared, "This reckless decision will threaten tens of thousands of American jobs and hurt our climate." The Solar Energy Industries Association, the industry's main trade group, claimed that the tariffs would increase energy costs and put "48,000 to 63,000 American solar industry workers out of a job this year." That would wipe out up to one-quarter of the solar industry's 250,000 jobs.

Devastating. And it never happened.

In fact, the job decline in solar energy was fewer than 8,000 jobs, or just 3.2 percent, in the year after the new 30 percent tariff went into place. The year after that, in 2019, the solar industry was back up to 249,983 jobs, 17 shy of the 2017 level. So the Trump tariffs led to a net decline in solar energy–related jobs that amounted to 0.1 percent. It was a small price to pay to begin taking back some rope in our tug-of-war with the Chinese. These numbers come from the Solar Energy Industries Association, whose website trumpets, "Solar Industry Growing at a Record Pace." It claims that "Solar energy in the United States is booming."

Yet in December 2019, the very same group put out a doomsday report—which the lib left-wing media called a study—timed for release the week the tariffs were to go up for a federal review. Suddenly the tariffs were devastating again. According to Reuters on December 3, 2019:

> The U.S. solar industry warned on Tuesday that the Trump administration's tariffs on imported panels will cost the United States 62,000 jobs and $19 billion (£14.81 billion) in investment, an estimate the White House dismissed as "fake news."

It is utterly confounding to me that in the age of social media, the internet, and a wealth of free information accessible to more people on Earth than ever before, it has become harder than ever to divine what is the objective truth, which facts are slanted and which can be trusted, what is news and what is fake news, what is real.

The mainstream media were supposed to be the arbiters of truth, and back in the days of Walter Cronkite, maybe they were almost that. Especially since the election of President Trump, though, they have become part of the problem. A big part.

And what became of Whirlpool and the washing machine industry? The media played up the problems and decidedly mixed results, yet on balance the new tariffs achieved some key intended effects. Imports of foreign-made washing machines to the United States fell by 54 percent from 2017, before the tariff, to mid-2019. Before the tariff, imported machines had come in at a rate of 350,000 washers a month, or 4.2 million per year; by mid-2019, that was down to 160,000 a month, the *Wall Street Journal* reported.

The paper stopped short of admitting it directly, but the new tariff actually *created* hundreds of new jobs. Whirlpool hired two hundred more workers for a washer plant in Ohio in anticipation of more sales because of a decline in imports. That also meant more business from Whirlpool for an Ohio supplier of counterbalances for washer door hinges, the paper reported.

Moreover, Samsung and LG opened new plants in the United States directly because of the threat of the tariffs and the imposition of them thereafter. When President Trump first slapped the new tax on the imported white goods, LG instantly threatened job losses; didn't happen. Rather than retaliate and scrap their plans, LG and Samsung both moved forward—in part because the new Trump tariffs had made it too costly to source all of their US supply from their overseas plants. They still wanted access to our market.

In January 2018, just as President Trump was announcing the new tariffs, Samsung started making washers at an old Caterpillar plant in

South Carolina, the *Journal* reported. The new plant employs more than six hundred workers and was producing a thousand washing machines a day by the spring of 2019. It made better sense to make more boxes here. Score another one for President Trump.

One drawback: the prices of washers and dryers jumped in the months after the new 20 percent surtax took hold. Purchases of washing machines slowed because of the higher prices. That made for more negative headlines about the damage done by the Trump tariffs. Yet a year later, the price hike had faded. In a story in the *Times* on January 25, 2019, the reporter noted that although "laundry equipment inflation" had peaked at 20 percent in July 2018, by December it was down to just 1.6 percent.

It was a rather ridiculous overabundance of *sturm und drang* over the Trump tariffs on solar panels and washing machines. The two cases amounted to a gnat on an elephant's backside in the context of the $20 trillion in commerce that takes place in a year in the United States. They may fall short of doing much to help rebuild US manufacturing might in the solar business and white goods.

Nor was that the real intent of the president's laying down the new Trump tariffs. Their importance lay in their direction and milestones. For the first time in his new administration, President Trump had found a way to tax imports from China and elsewhere in Asia, swiftly and free of interference from the World Trade Organization, the Democrats in Congress, or his Republican colleagues in the Senate.

On his first foray into tariffs, President Trump had made progress in debunking the establishment canon that had dominated US policy for decades. He had defied the battering and pleading of all the special interests feeding at the free-trade trough: the gray-head leaders of his own party who had never stepped up and their Dem counterparts; big labor *and* big business; economists in government, corporations, and academia.

Virtually everyone said it was unwise to try and impossible to achieve—except a few loud voices from people who knew the destruction going on and kept yelling about it because no one was listening to them. They included Peter Navarro, Michael Pillsbury, Gordon Chang, and myself. And Donald Trump.

We bore the curse of a free-trade Cassandra. The priestess of Apollo from Greek mythology often is misrepresented as an ancient Chicken Little who wrongly thought the sky was falling. Her real curse was that she foresaw the truth that we were doomed and no one would believe her.

President Trump had taken an important first step in penalizing China directly for its underhanded tactics and predatory pricing. And he was just getting started.

CHAPTER FIVE

THE TRUMP TARIFFS

S ix weeks after President Trump imposed tariffs on solar panels and washer-dryers, he turned to imports of steel and aluminum from around the world. The real target was China, the largest steel producer in the world. And so it was that at 7:12 a.m. Eastern Time on Thursday morning, March 1, 2018, the United States, the richest and most powerful nation on earth, launched a trade war against the biggest nation on Earth.

On Twitter:

> Our Steel and Aluminum industries (and many others) have been decimated by decades of unfair trade and bad policy with countries from around the world. We must not let our country, companies, and workers be taken advantage of any longer. We want free, fair, and SMART TRADE!

The president's opening tweet surprised even some of his advisors, who were divided over whether he should go through with the tariff threats he had been making from the moment he started running for office. It set off a reaction that was swift and fearful. A global trade war might break out, said just about everyone, and other nations would retaliate against us and starve the US economy. That would shut down our growth and clobber the stock market. All because of Donald Trump.

A couple of hours later, trading commenced on the New York Stock Exchange. The news sent the Dow Jones Industrial Average down 500 points, more than 2 percent, in an instant. Stocks dropped even more for carmakers and other giant customers of steel and aluminum. Five hundred points was a big deal back in the day, before the Wuhan crisis and a few *daily* swings of more than a thousand points.

Fears of a trade war poured in from officials around the world, amplified by the media. The phrase "saber rattling" has been in our lexicon for more than a hundred years, and the reaction of our trading partners shows why. In Europe, the chairman of the European Parliament's International Trade Committee, Bernd Lange, intoned, "With this, the declaration of war has arrived."

The president of the European Commission, Jean-Claude Juncker, blasted away: "None of this is reasonable, but reason is a sentiment that's very unevenly distributed in the world. . . . I can't see how this isn't part of war-like behavior." In a statement, he fulminated, "We will not sit idly while our industry is hit with unfair measures that put thousands of European jobs at risk."

Australia's trade minister, Simon Birmingham, warned that the new US tariffs could create a global recession, "and we know the consequent impact of that."

In the United States, even leading members of President Trump's own party joined the breakdown over his tariffs. On *Roll Call*, a venerable paper covering Washington since 1955, the online headline said it all: "GOP Reaction to Trump Tariffs Is Fast, Furious and Negative," with the subhead "Republicans fret about retaliatory action, effect on agricultural trade." They included Speaker of the House Paul Ryan and Senators Ted Cruz of Texas, Lindsey Graham of South Carolina, Orrin Hatch of Utah, and the Trump traitor who later replaced him, Mitt Romney.

Ryan's office put out an official putdown: "We are extremely worried about the consequences of a trade war and are urging the White House not to advance with this plan." On *Face the Nation*, Graham addressed the president through the TV screen rather than in a private face-to-face: "You're punishing the American taxpayers, and you are making a huge mistake." Republican Kevin Brady of Texas, the chairman of the House Ways and Means Committee, had visited the White House twice the previous week to urge the administration to hold off.

Shockingly, President Trump got more support from the Democrats, albeit in a backhanded way. In coming days, Chuck Schumer of New York, the Senate minority leader at the time, would say that the president was "doing the right thing when it comes to China"—though he also predicted that the tariffs would "cause more damage to key allies and our domestic industries" and slammed "the haphazard way these tariffs were put together."

Nancy Pelosi labeled the move "merely a start" and called on the administration to "show the moral courage . . . to advance human rights in China and Tibet. If we do not speak out for human rights in China because of economic concerns, then we lose all moral authority to talk about human rights in any other place in the world."

Big business also weighed in against President Trump's new plan. The US Chamber of Commerce, a week into the steel controversy, put out a statement saying it was "very concerned about the increasing prospects of a trade war." The big-business lobby urged the president to refrain from his plan, then lectured him, "Alienating our strongest global allies amid high-stakes trade negotiations is not the path to long-term American leadership." The Chamber also trod upon Trump turf, tweeting from its Twitter account, @USChamber, on March 2:

> Trade with Canada and Mexico supports 14 million American jobs. Learn why #NAFTA is crucial to our economy.

It then offered a clickable link for the wonkiest among us.

The tariff backlash went beyond big business. A Minnesota tool-maker told the *Wall Street Journal*, "It's going to be expensive. . . . All of it will impact the consumer." On Twitter, in the president's feed of replies, an early response came from a Trump voter with all of thirty-one followers, Steve Thompson (@capnsteveo007):

> I voted for you but with one stroke of your pen, you ruined everything. The company I work for and invest in makes products out of steel. The local mills are already increasing prices and rationing deliveries. We will lay people off. You screwed millions to help a few.

In that sea of opposition here at home, President Trump did get a bit of support from two cable network commentators. CNBC.com posted a story the first evening on what Jim Cramer, the Crazy Eddie of stock-show hosts on CNBC's *Mad Money*, had to say about the new tariffs. Its headline ran, "Worries About the Trump Tariff Are Far Too Extreme." A few days later, the stock market rose nicely, and @JimCramer tweeted:

> If this is your idea of a bear attack . . . we clearly need more tariffs.

And there was this wise tweet from another cable host:

President Trump means to end dumb government and dumb trade
policies that sent millions of American jobs abroad and cost the
American economy trillions of dollars over the past half century.

Full disclosure: That one came from yours truly.

It was as if mass hysteria and Munchausen-by-proxy syndrome had
overtaken the global trade establishment—and it had. So many people in high places were apoplectic and panic stricken, yet it was China
that was going to take any hit. Why the overdose of doomsday fear?

One reason is that the new Trump tariffs posed a threat to the very
existence of the old bureaucracy that had grown up around world
trade since the World Trade Organization had been formed in 1995.
President Trump was disrupting the fiefs of the regulators: staff, tran-
scribers, analysts, expert witnesses, examiners, lawyers, judges, and
more, not just at the WTO itself but at EU and Asian trade organizations and at the governments of the WTO's 164 nation members.

The WTO is a massive multiheaded hydra that harks back to a
much slower world. Today it illustrates the pointlessness of international organizations hewing to their antiquated and eroded way of
doing things. It opened for business twenty-five years ago and still op-
erates at the pace of that time. The world of the internet, smartphones,
and AI hurtles forward at light speed, yet the WTO remains the Rip
van Winkle of trade: slumbering, blissfully unaware of how the world
has changed.

The WTO can take years to resolve a dispute between two countries over a product that had a shelf life of a year. It is guided by a
seething anti-American resentment and a passion for protocol and
red tape.

Thus, in the blinding speed of the Trump Century, the WTO
is a nice thing to have—but unequal to the tasks before it. On the
day President Trump imposed the new levies on the worldwide steel

business, while the rest of the world huffed and puffed to great effect, the best outrage the WTO could summon was to say that it was "clearly concerned" and "the potential for escalation is real."

The president lacked any patience for the WTO's sluggish metabolism and protracted resolution process. He moved a lot faster than the cumbersome body—how can 164 equal members of any group agree on anything decisive and dynamic? They cannot, and the new president lacked all tolerance for it. He was going his own way, as he always had.

In China, reaction to the direct assault on its world domination of metals production was split. Government information agents plied a bifurcated strategy of bad cop, good cop. Bad cop: the China Iron and Steel Association, whose vice secretary-general issued a statement calling the new tariffs "extremely stupid" and "a desperate attempt by Trump to pander to his voters." Good cop: China's Foreign Ministry, which put out a tepid statement vowing to take "proper measures to safeguard our interests." That was a show of uncharacteristic restraint, especially when Chinese officials had to have been furious.

On the day before he imposed the metals tariffs, President Trump had announced the prospect of slapping surtaxes on $60 billion of other Chinese goods, including electronics. That on top of the tariffs on solar panels and washing machines he had announced on January 22. Now the president had chosen that day, March 1, to unveil tariffs on steel and aluminum—just hours before White House aides would host the top economic advisor to President Xi to discuss how to ease trade tensions. Awkward. The *Wall Street Journal* called his timing a "diplomatic jab."

The morning after all the tariff news, President Trump dug in, as he is wont to do, bypassing the corporate left-wing media mob to tell his 77 million Twitter followers why the new tariffs were necessary.

His first tariff tweet went out at 5:50 a.m., the second at 8:01 a.m., the third at 8:57 a.m.

When a country (USA) is losing many billions of dollars on trade with virtually every country it does business with, trade wars are good, and easy to win. Example, when we are down $100 billion with a certain country and they get cute, don't trade anymore—we win big. It's easy!

When a country Taxes our products coming in at, say, 50%, and we Tax the same product coming into our country at ZERO, not fair or smart. We will soon be starting RECIPROCAL TAXES so that we will charge the same thing as they charge us. $800 Billion Trade Deficit—have no choice!

We must protect our country and our workers. Our steel industry is in bad shape. IF YOU DON'T HAVE STEEL, YOU DON'T HAVE A COUNTRY!

In the days before Twitter, typing in ALL CAPS in an email or text was considered rude, equivalent to yelling. Our president has turned it into an art form. Each of the tweets garnered "likes" from a hundred thousand followers and got retweeted by twenty thousand, a viral chain that can spread exponentially to reach millions more people.

Three weeks later, on March 23, the US Customs and Border Protection agency started collecting duties on metals imports from around the world. The new tariffs, 25 percent on steel and 10 percent on aluminum, applied to imports of $50 billion in the metals in 2017. That was small beer in the context of the United States' $21 trillion GDP (before Wuhan). Then again, context stopped mattering long ago to the Fake News media and the multitudes of Trump bashers.

The critics of the Trump tariffs predicted retaliation from US trading partners, and they were right—and wrong. The counterpunch from overseas was more like shadow boxing, aimed at all show and doing as little damage as possible.

The European Union filed a complaint with the WTO, then waited more than two months to move ahead with new tariffs on all of $7.5 billion of US exports. A pittance. Canada set tariffs on up to $12.8 billion of imports of finished steel and aluminum from the United States. India imposed penalties of $200 million and delayed their taking effect for a full year. Teensy.

Certainly, those measures made for dramatic headlines. As it would turn out, though, very little of the much-hyped destruction happened, and even less impact fell on the American consumer, the engine driving more than two-thirds of the US economy. Inflation remained dormant, US unemployment continued heading down to all-time lows (before Wuhan), the world kept turning. GDP growth plugged along higher than the Fed's 2 percent expectation. Stocks would rise another 20 percent in the next year, before the coronavirus crash.

How could so many experts and policy makers get so many things so wrong about the Trump tariffs? Sage, well-educated, highly intelligent people with years and decades of experience. It boggles the mind.

The same day the tariffs dropped, CNBC asked Adam Posen, the president of the Peterson Institute for International Economics (PIIE), for his reaction. He replied, "This is just straight up stupid. This is fundamentally incompetent, corrupt, or misguided." Days later, a colleague, Chad P. Bown, also with PIIE, posted a takedown, saying that the president's powers to raise tariffs by "nontransparent procedures also raise concern over their potential for abuse, arbitrary decision-making, and corruption."

PIIE claims to be nonpartisan, but its board includes three Obama alumni (Treasury secretary, budget director, and national security advisor), an IMF official, and senior executives of IBM, Caterpillar, Morgan Stanley, Citigroup, PIMCO, BlackRock, Aetna, and Aflac.

In other words, the lib-corporate-elite establishment that had gotten us there.

The author of the article, a member of Obama's Council of Economic Advisers, was attacking the element that gave the new tariffs a chance to work. The president would have wide latitude to levy tariffs at will. That would give him an edge: speed.

Global trade disputes at the WTO are unworkable, especially in high-tech. It is driven by a doubling of computing power every eighteen months, for the past fifty-five years and counting. Business and national economies move faster than ever, the entire world moves faster, as if we were in hyperdrive. Especially in the time of Trump.

It is a nonstarter that Donald Trump would trifle with the WTO's nettlesome procedures. Going to Congress was a nonstarter, too many enemies on both sides of the aisle. It is just as deliberative as the WTO and even more hostile. So he skipped that rigmarole, jumping past the WTO and everyone else.

The administration's resulting approach, invoking national security concerns and obscure sections of laws that had been left behind long ago (see chapter 4), gave President Trump more room to maneuver.

Technically, the new tariffs on steel were aimed at nations other than China. Though it is the world's largest producer, with half the worldwide market, it had left the US market in early 2016, after the Commerce Department had imposed an astronomical 265.79 percent tariff on steel imported from China for trade violations. That is not a typo; the feds were that precise.

That effort flopped when China simply rerouted production and shipping to other countries. The Trump administration's aiming beyond China, as a way to target China, was a reason cited for Democratic opposition to the Trump plan. It is a weak excuse. How is the

United States supposed to directly block China from producing whatever it wants to produce, in whatever quantities it wants?

Taxing imports from everywhere else (save for the allies that get waivers from the president), when lots of their steel also comes from China, was a smart way to target our adversary.

China has more than a thousand steel producers stamping out more supply than it requires domestically. It offloads the surplus onto the global market at low prices that depress prices for everyone else. That was the essence of the Trump administration's argument for why the new tariffs should apply to all foreign steel imports. It was a crafty side entrance to go after the largest producer of steel in the world.

Beijing first targeted the global steel market twenty years ago as one of the industries it hoped to dominate. In the United States, an industry that hopes to dominate global competition must wage this crusade on its own. The United States lacks an industrial policy that targets cutting-edge industries and lends the largesse of government financing to help its home-turf companies defeat overseas rivals.

Donald Trump has always thought that was clueless of us.

We hear complaints in the United States about corporate welfare and crony capitalism, but what the Chinese do is beyond the pale. Illegal trade practices, government-subsidized production that artificially lowers costs, export rebates and quotas, an intentionally undervalued currency, subsidized financing, weak regulations on environmental, labor, and safety issues: these enable China to underprice the rest of the world.

This slate of predatory practices is the real reason behind China's incredible growth story of the past three decades. Without the government's heavy help, Chinese manufacturers would never have risen up so fast and so hugely.

In the year 2000, China produced almost one hundred thirty million tons of steel, 15.5 percent of the global market. Its number two rival, India, held 3 percent of the market. China produced almost five times as much steel as its closest rival—and ten times as much as US steel producers.

A decade later, China was producing almost 600 million tons, up almost fivefold in ten years. Its worldwide share had more than tripled to 47 percent. India's share had risen to 5.1 percent. China was now producing *eight* times as much steel as its number two rival.

In the United States, the steel industry was getting worried about its ability to compete with cheap Chinese imports. In 2009, the Congressional Research Service published a report on the issue, presciently noting industry worries about the coming threat:

> *Although industry statistics indicate that the Chinese steel industry is not export-oriented, its consistently high output keeps U.S. steelmakers concerned that excess Chinese steel might overwhelm the global market once domestic demand is adequately met. These concerns become increasingly acute as the United States and the rest of the world are in the middle of a slow recovery from the economic recession started in December 2007.*

The Great Recession prompted others to cut back and hunker down—and this is when China doubled down and started producing even more steel. After the global meltdown in 2008, China set a stimulus program of more than half a trillion dollars, or 11 percent of China's annual GDP. That is equivalent to spending $1.55 trillion in China today. That helped fund a new push in steel production.

Part of the Chinese government stimulus program entailed underwriting the construction of hundreds of new steel factories. Overall, production rose sevenfold from 2000 to 2013. By 2018, Chinese production had grown another 60 percent since 2009, to more than 900 million tons of steel per year, with a 51.3 percent share of the global market. In the same period, India's share had tripled to 15.5 percent.

In the five-year period up to year-end 2016, the price of steel made in China crashed by 50 percent.

The fat 265 percent tariffs the United States imposed on Chinese steel in 2016 prompted China to expand factories elsewhere—same producers, different address—to sidestep the US levies. In short order, Chinese steel producers used cheap government financing to build plants in India, Indonesia, Malaysia, Serbia, Brazil—and Texas. If they had done a lot more in Texas and the United States, our relations might have fared better over the years.

This carpet-bagging mobility is very much like the moves made by the washer-dryer businesses of Samsung and LG when Whirlpool was pursuing a 2012 complaint at the International Trade Commission. (See chapter 4.) We tariffed white-goods imports from South Korea and Mexico, so Samsung and LG hopscotched to China, then to Vietnam and Thailand. After which Whirlpool gave up and turned to Trump.

The WTO's long-winded process is feckless in countering China's aggressive moves. The administration's new push looked more promising. On March 1, 2018, as President Trump announced the new tariffs on all imports of steel and aluminum, he went into fierce deal-making mode, staging a series of threats, feints, bluffs, and actions.

Initially, the tariff target was going to be $50 billion a year in imports from China. It was a bluff. A week later, Trump reduced the target by $15 billion by granting exemptions to our partners in Canada and Mexico. Though it was more of an administrative procedure, he held court for the Fake Media—making use of them even as he pilloried them—at a ceremony where he put his giant Sharpie signature on two proclamations stating that the tariffs would take effect on March 23.

The day before the Trump tariffs were to kick in, President Trump issued two more waivers, one for each metal, freeing the European Union, South Korea, Brazil, Australia, and Argentina. That meant that more than 60 percent of the steel and aluminum imports first included in the tariffs now had waivers.

President Trump was closing the cordon around his real quarry: China.

The late-breaking waivers gave our trading partners a bit of relief, and the president had just won their attention and maybe a little gratitude—for giving in on a thing he had imposed on them. In the coming months, President Trump would treat the tariffs and the bestowal of waivers, exemptions, and delays as a kind of patronage system for rewarding the United States' friends and throwing shade at our antagonists, China chief among them.

In July 2018, President Trump announced an investigation into whether tariffs were necessary for imports of cars made in Europe. Might they pose a threat to US national security, given how much steel and aluminum are used by their manufacturers? A month later, he delayed the tariffs, as European negotiators were about to sit down with his team to talk new terms. He also got started on throwing out NAFTA and pushing Canada and Mexico into signing a better deal of his administration's design: USMCA.

A few months later, in September 2018, President Trump and Japan announced plans to strike a bilateral accord, though Japan was a leading advocate of the Trans-Pacific Partnership, which Trump had rejected. Ultimately, the two nations would sign a new bilateral accord that called for the Japanese to let 90 percent of the US-made goods they buy arrive without any tariffs, up from less than 40 percent previously. It covers $14 billion a year in US sales to Japan.

Meanwhile, President Trump kept threatening to throw still more tariffs at China. He had started with washing machines and solar panels. Now he had the 10 percent tariffs on aluminum and 25 percent tariffs on steel. Later, he would extend the levies to Chinese-made electronics, machinery, and components.

At first, the threat was to apply the tariffs to $60 billion of imports; then he reduced it to $50 billion. Thereafter, President Trump first said that he would impose 25 percent tariffs on $100 billion of Chinese goods, only to reduce the rate in July 2018 to 10 percent—but he *doubled* the total value of goods covered to $200 billion. They included flat-panel displays, smartphone components, luggage, tires, baseball gloves, tuna, and salmon, as the *Wall Street Journal* reported. The paper added:

> *The early reaction out of Beijing was scornful. "There is a proverb in the West, 'like a bull in a China shop,'" said Li Chenggang, an assistant minister of commerce, at a conference in Beijing. "The U.S. approach undermines the process of globalization and undermines the trade order."*

A Chinese commerce minister quoting a "proverb in the West." That has to be a first. Usually, Westerners quote Chinese proverbs, although "bull in a *china* shop" is more of a nineteenth-century American aphorism.

In June 2018, China retaliated by imposing its own tariffs on almost $35 billion of US farm products, aircraft, and other wares and said it would match any more Trump tariffs dollar for dollar. To which the president responded that if China adopted new levies, he would add even *more* China exports to his ban: an extra $250 billion more. That would have more than doubled the tariffs on China to $450 billion! As the *New York Times* noted, a little startled, it was "a sum nearly as large as the total value of goods China sent the United States last year, which was $505.6 billion."

And then another Trump twist, as reported in the *Journal* on Sept. 13:

> *The Trump administration is giving Beijing another chance to try to stave off new tariffs on $200 billion in Chinese exports, asking top officials for a fresh round of trade talks later this month. . . .*
>
> *The invitation from Treasury secretary Steven Mnuchin comes as some Trump officials said they sense a new vulnerability—and possibly more flexibility—among Chinese officials pressured by U.S. tariffs imposed earlier this year and threats for more.*

For two decades, Chinese leaders had told their trading rivals to go pound sand, and nobody had stood up to them. Now the most influential business newspaper in the world was reporting that the president was "giving Beijing another chance to try to stave off tariffs on $200 billion in Chinese exports." It was a threat to China that had never existed before. Now it existed only because President Trump had created it. It was rather resourceful and shrewd of him.

Just eighteen months into his presidency, US Customs agents were levying the new Trump tariffs on foreign imports as they arrived at US ports. The new surtaxes were levied on China and on the rest of the world to cover the subsidized, artificially cheap steel, solar panels, and other products made in China and shipped elsewhere before arriving in the United States.

The heads of the World Trade Organization would have gone into cardiac arrest on maneuvers like the ones President Trump was pulling. The WTO website provides a flowchart mapping its "dispute settlement process." The long journey entails two different approaches, three main stages, and sixteen major steps. Plus appeals. Cases can linger in purgatory for years.

The WTO's sclerotic process starts with sixty days of "consultations" before establishing a panel to hear the case. Then members have up to twenty days to figure out "terms of reference," plus up to thirty days more to hold three meetings. Then comes an expert panel review, an interim stage review, and a review meeting with the panel.

This is followed by a report issued to the parties, then a report issued to the Dispute Settlement Body (DSB). From there, an appeal can consume ninety days more, followed by final adoption of the report by the DSB in sixty days, unless the loser appeals—which adds thirty days on top of that. Followed by implementation of the report's rulings, and another ninety days if that leads to a further dispute. Plus thirty days more for "retaliation," if the parties are unable to agree on compensation.

Gasping for air yet? The WTO says that once a case's panel is established (within sixty days of filing a trade complaint), it should take nine months for the panel to issue a final report to the DSB, before appeals, penalties, and the like.

At minimum, the WTO process requires 410 days, more than thirteen months. Even that estimate is unrealistically hopeful. The WTO website lists hundreds of musty, old cases involving parties that must have moved on with their lives long ago.

Example: Case No. DS 452 (the DS stands for "Dispute Settlement") is a complaint against China filed by Mexico: "China—Measures Relating to the Production and Exportation of Apparel and Textile Products." Mexico filed its complaint on October 15, 2012. "Current status: in consultations." Eight years of jousting with the Chinese over apparel that went out of style long ago.

After the Trump tariffs went into place, nine complaints were filed at the WTO by China, the European Union, Mexico, Norway, Canada, Russia, India, Switzerland, and Turkey. Canada and Mexico withdrew their cases after agreeing to a new Trump trade pact, USMCA. The

EU complaint took twenty months to wind its way through the WTO python before lawyers got a chance to argue the case.

On November 4, 2019, officials for the European Union and the United States presented their arguments to a WTO panel of three "adjudicators," from Uruguay, Chile, and the Philippines. Now the parties must wait another full year for the WTO to make a ruling. The WTO may leave the case lingering until after the presidential election. No wonder President Trump chose to take a faster alternate route.

In 2018, Japan instantly threatened retaliation and cited US imports worth $409 million. Yet Japan would end up signing a new trade deal with the US effective January 2020. Previously, only 36 percent of US food and agricultural exports to Japan had been tariff free. Now, under the Trump deal, that duty-free portion rose to 90 percent. It is a small annual number, $15 billion total, but it is a start—and a total reversal of the way trade has worked for an awfully long time.

Meanwhile, Russia also moved against the new Trump plan. Three weeks after the announcement, two subsidiaries of the Russian steel giant PAO Severstal filed a lawsuit against the United States seeking an injunction from the US Court of International Trade. That case may be moot: a year later, in March 2019, the court ruled in a separate case that the Trump tariffs are constitutional.

Even China, the most formidable rival we have, struck back with only a glancing blow. It declared tariffs equivalent to what the United States had imposed. For all the economic damage the doubters had feared, it would be minuscule in comparison to the devastation wrought by the Wuhan pandemic.

Yet we were told, with such great passion and certainty, that the world would catch on fire because of the Trump tariffs. There was no room for debate. One reason is that so much was at stake for the US ruling class, which had risen to success and wealth on the way free trade had always worked. People hate change—especially when they have everything to lose.

After two decades of taking a beating in world trade and losing millions of manufacturing jobs, the United States had lost its moxie. We had developed an inferiority complex. We could manage the United States' decline, but we could do nothing to reverse it.

That was defeatist and milquetoast, given the great people, natural resources, enterprise, and innovative spirit of our country. We have won more than half of all Nobel Prizes. We have the most fertile farmland in the world. China puts its ethnic minorities into reeducation camps, while Americans happily vote theirs into public office. There's absolutely no reason to settle for second place (or worse).

President Trump possessed more faith and confidence in the strength of America and our people than everyone else did. The fretful opponents of the new tariff program forgot a key advantage the United States has in trade wars: everywhere else in the world, people need us more than we need them. Even the Chinese. Our people are among the richest on Earth, with lots of disposable income to spend. Our economy has the most innovation, the most advanced technology, the most robust infrastructure, and the strongest rule of law, all prerequisites for growth.

This means we can withstand a trade conflict longer than our rivals can, while enduring less damage. The president often says the tariffs are a tax on the countries that have taken advantage of our kindness and cluelessness. Impose new taxes to reclaim some of the hundreds of billions of dollars the United States has let them drain from our accounts year after year. If they retaliate, let both sides hold on tight and see what emerges. Ultimately the United States can win.

Even after China's torrid growth for thirty years, the United States is still the largest economy on the planet, a title it has held since 1871.

China is gaining on us every year, but by 2019 US GDP was still 50 percent greater than China's: $21.44 trillion versus $14.14 trillion, according to the International Monetary Fund. Our per capita GDP was more than six times that of China ($64,767 versus $10,153), and our national per capita income was almost triple China's when measured in purchasing power parity, or PPP. That is considered the best metric for comparing the incomes of two very different nations in apples-to-apples terms.

Those impressive spreads are woefully short of telling the real picture. As the United States grew up, faith in our economic freedom flourished in the fertile soil of American prosperity. The size of our continent-spanning nation gave us the largest free-trade zone in the world for most of our history. The diversity of climates and conditions across the continent made us less dependent than other nations on foreign trade. We have the great fortune of being endowed with abundant natural resources that seem to show up just when we need them, almost providentially.

In the 1970s and '80s, our future looked bleak and beholden to the Mideast oil producers that supplied us but disliked us. The rise of hydraulic fracking in the United States revealed that we had enough oil and gas to become energy independent and stay that way as far as the eye can see. In the 2020s, I would bet, we will see a vaccine and effective treatments to turn the Wuhan virus into an almost forgotten, old-world malady like the measles. And I hope US companies will be at the forefront of that achievement.

The greatest threat to the economic freedom so deeply entwined with our moral values comes from within rather than from abroad. It comes from the callousness of a political class in Washington that taught itself to be indifferent to the unemployed and unmoved by the stagnation of wages of working men and women.

Someone had to break this downward spiral in the American spirit.

For President Trump, the new tariffs he had just imposed on steel and aluminum, worldwide and in one fell swoop, were another step toward that goal. He always said the tariffs were a tax to force the nations that had been fleecing us to start sending hundreds of billions of dollars back to the United States. The more important underlying objective, however, was to force China to sit down at the bargaining table. Then the president and his advisors could start negotiating reforms and restrictions that can rein in that country for decades to come.

It was classic Trump style. First, start with some kind of gauntlet and throw it down for your adversary. Then, once you have his attention, seek negotiations aimed at forcing still more concessions. Even in New York, this is an awful lot of chutzpah. Yet, ultimately, the Trump tariffs would achieve the impossible, forcing China to the negotiating table.

The biggest trade violator in history would end up admitting to the world that some trading terms were unfair and in need of fixing. Eureka! Without the prodding and threats of President Trump, China would have had no reason to make any change. Everyone else gave China free rein—and a free reign—to be a selfish, irresponsible actor in global economic affairs.

Only President Trump stood up, pushed back, and said "Enough!" For the rest of the Trump Century, no president will be able to get away with hiding behind the cloak of the free-trade banner and weak and soporific regulatory bodies such as the World Trade Organization. There was a better way to proceed, and President Trump had shown us that.

Let the haters doubt and fret and sling fear across the land. President Trump was already plotting his next major initiative. It was going to be "yuge," as he might say. He wanted to prod the Chinese to sit down for the first time ever with an eye toward constraining their lethal and unfair practices.

No longer was the topic of talks what everyone else could do to help China. Now it was going to be about what China could do to square up with the United States.

The president already had a big number in mind to sell his next program. He had just slapped Big China right in the kisser. Now he wanted to talk our enemy into buying more American-made goods. He even had a number in mind: maybe $100 billion, maybe $200 billion. Tariffs be damned. That's up next.

CHAPTER SIX

CHINA BOWS

It took a while, and for President Trump, the long wait must have been worth it. Clock it in at just shy of two years—and three decades of his life.

On the morning of January 15, 2020, three years into his presidency, President Trump was standing in the East Room of the White House before an audience of more than two hundred members of the Washington elite. He was about to sign phase one of a "yuge" trade deal with China. The vice premier of China was in attendance, and President Xi was present by flat screen, his image beamed in live from Beijing.

That happened less than two years after the president had declared a trade war on China; after thirty years of unfair trade terms that had drained trillions of dollars out of the US economy and funneled most of them to our most formidable global rival. The Trump tariffs were raking in billions of dollars as a result. The more important upside was that, for the first time in US-China trade relations, America was pushing back, and China was about to give the bow.

I was there in a front-row seat, and I feel proud to say it.

It was a moment of historic triumph for President Trump, though his adversaries and the Fake News would fail to acknowledge it. He had prevailed once more over the doubters and the haters and pixel pundits who said he would fail to bring China to heel. They said his tariffs, an idea he had harbored and talked about for years, would wreck the US economy and spark destruction and retaliation worldwide.

Instead, the Trump tariffs had prodded China to the table. And pushed it into making concessions and alterations in trade terms that were unfathomable before we hurtled into the Trump Century.

At 11:51 a.m., as the news networks covered the ceremony live, President Trump opened with a scripted and suitably presidential pronouncement appropriate to the occasion, reading from the teleprompter:

Today, we take a momentous step—one that has never been taken before with China—toward a future of fair and reciprocal trade, as we sign phase one of the historic trade deal between the United States and China. Together, we are righting the wrongs of the past and delivering a future of economic justice and security for American workers, farmers, and families.

That would be the last time he consulted the teleprompter for quite a while, as he held forth in an extraordinary one-man show that was quintessential Trump. On display, and live on cable, were the qualities that charm and invigorate his supporters—and that outrage and mortify the Never Trumpers.

After the networks went live and President Trump started the proceedings, he had a captive audience, and he used it to the utmost. He went wildly off script. That gave the networks no clue as to what he might say and when he might reveal the details of the trade deal, which made it harder for them to cut away.

For the next hour and fifteen minutes, he ad-libbed, riffed, and joked, mixing equal parts candidate at a rally, cheerleader for business, and Jimmy Two Times from *Goodfellas*. ("I'm gonna go get the paper, get the papers.") Reveling in the moment, he paid tribute to sixty important personalities in attendance, introducing them one by one. Advisors, politicians, donors, business chiefs, supporters, and other hangers-on were among the throngs watching the show—so many that the event had had to be moved from the Oval Office to the larger East Room.

In between the intros and the anecdotes, the president dropped bombs on his assailants. He took shots at Cryin' Chuck Schumer and the Democrats for a procedural impeachment vote under way in the Senate at the very same moment. He got in a few more digs in his feud with the chairman of the Federal Reserve for keeping interest rates too high. The president had major news to release, and he was making everyone wait. You don't have to wait. Here are the highlights.

The new trade accord was something Joe Biden might have called a Big Effin' Deal. It committed the Chinese to buying $200 billion in US goods in the next two years. China also consented to crackdown provisions on makers of pirated and counterfeit goods. It accepted restrictions on currency devaluation, a favorite tool for making its exports to the United States cheaper than our exports to China.

Even though China made those precedent-setting concessions, the Trump tariffs were *staying in place*. That is remarkable. The whole reason to make concessions in a negotiation is to get the other side to do so in return. In the era of Trump, some trade deals *can* be a one-way street—in our favor for a change.

The new Trump trade package aims at bigger targets than the familiar complaints about stolen IP such as pirated software and unauthorized DVDs of Hollywood movies. It entails a big win for US companies wanting to set up shop in China. For years US firms were required by the government to form joint ventures with Chinese firms, many of which were linked to the ruling Communist regime— and even hand them majority ownership. If US multinationals had the option to fight back and refuse, they chose against exercising it. They were too enamored of the promise of the China market.

As President Trump explained to his listeners in the East Room:

We never even had a deal with China. In all fairness, I don't blame China. I blame the people that stood here before me. I don't blame China. I told that to President Xi. I was in Beijing, making a speech, saying how they're ripping us off. And guess what? He wasn't too happy. I looked at him. I said, "He's not happy." And I said, "I better change the speech quickly."

So I said, "I don't blame him. I blame our Presidents." And I'm right. We should've done the same thing to them, but we didn't. We didn't. We never had a deal with them. They'd do whatever they wanted.

And then back to the teleprompter: "With this signing, we mark more than just an agreement: we mark a sea change in international trade."

The president and China had been on a collision course for three decades. In that time, China had transformed itself from a backward nation, starkly poor and rural, into a powerful, teeming economic miracle. That rise had been fueled by the kinds of unfair and brutal tactics that Trump had decried from the start of his career.

Rapaciously and with impunity, China had gotten away with leveraging the massive size of its market as a lure for US multinationals. If you want access to our hordes of upwardly mobile consumers, it said, you have to play by our onerous rules: establish joint ownership, provide us with access to proprietary technology.

Companies succumbed to that and complained to US officials that China was coercing them into surrendering valuable intellectual property and technology. Fearing reprisals, however, those complaints had been made quietly. It was the biggest complaint from high-tech companies in the early days of the Trump administration—but not one of those tech titans would go public with its gripes for fear of losing access to China's labor force and consumers.

Yet China never truly opened its economy to unfettered outside investment and robust competition. It remains closed to outsiders in many ways. It cannot fathom truly fair trade because it never has even been asked to abide by it.

This means that China's economy and its growth depend far more on exploitation of the global trade system and unfair trade terms than even many of its critics believe. China hopes to grow into the largest economy in the world and supplant the United States as the most powerful country in the world. But rather than achieving this with innovation, creativity, and diligence, it relies on its special brand of predatory mercantilism: cheating, lying, coercion, deception, and theft of intellectual property.

The new Trump treaty aims beyond the familiar complaints about stolen IP, such as pirated software, unauthorized DVDs of Hollywood movies, or knockoff T-shirts bearing the likenesses of Mickey Mouse or Disney princesses. US firms were forced to form joint ventures with Chinese firms, many of which were linked to the ruling Communist regime. The Chinese "partner" would own the majority stake in the joint venture, and much of the manufacturing would be done in China.

US businesses had seen repeated instances of Chinese companies' obtaining their trade secrets and their technology, and launching rival products wholly owned by Chinese firms—in competition with the US partners that had been forced to ally with them. Yet US companies never banded together in a united front to combat the Chinese offenses, and one reason is the cutthroat competition of capitalism. Each one feared to leave the market to its rivals if it dared object and got thrown out of China.

The system was as simple as it was devious: if a US company wanted access to China's market of 1.4 billion consumers and its growing middle class, it had to sign up with a Chinese partner, hand technology over to the partner for manufacturing in China, and then keep quiet when it was stolen and used by competitors. It had to hand over the tech or get frozen out, while its rivals sold cars, phones, software, and microchips to Chinese consumers. China was using the competitive nature of the United States' capital economy to prey on American know-how.

While smaller US businesses had often found that their intellectual property was stolen outright by Chinese companies, China also had succeeded in coercing even the largest American firms. General Motors and Microsoft had bowed to forming joint ventures in China.

A survey of the CNBC Global CFO Council, which represents large public and private companies in the world with a combined market value of almost $5 trillion, found that almost one-third of them said they had had intellectual property stolen by Chinese firms. The rest were uncertain.

China had also adopted a cybersecurity law that required US tech companies to store their Chinese users' data in China and provide the Chinese government with encryption keys. US firms, including

Google and Amazon, were at risk of aiding the Chinese surveillance state in its oppression of the Chinese people, while also providing China with access to their trade secrets and source codes that would fuel copycat rivals.

For China, another advantage of the joint-venture system was that it anchored manufacturing in China, even as the country grew wealthier and its workers' wages rose. In a truly free market, US and European firms would react to rising labor costs in China by moving production into countries with lower labor costs, such as nearby Vietnam. Some have done that. But China required manufacturing to remain in China and under the purview of the Chinese-controlled joint venture as the price of access to China's market. So the jobs stayed in China.

Get that? We did nothing to stop US companies from shipping jobs and production overseas; let free trade blossom and let the free market dictate those flows, our leaders said. And we lost several million manufacturing jobs. Then when China started to rise up, it passed rules blocking companies from fleeing its land for cheaper supply chains across its borders.

The result was that China became the Hotel California of manufacturing jobs: you can check out any time you like, but you can never leave.

This created imbalances that further incentivized US multinationals to set up shop in China and hand over the keys to the locals. If a smartphone maker such as Apple were to build phones exclusively in the United States, China would outright deny the company access to Chinese consumers or charge a tariff. If that company moves production to China, it faces no tariff on exporting the phones back to the United States and no tariff to sell them to China. And if it moves its production to Vietnam in search of cheaper labor, it faces the Chinese tariffs all over again.

China's rise to manufacturing dominance had nothing to do, really, with free trade and cheaper labor costs. Its success owes to brutal, predatory mercantilism.

Nearly all economists and economic policy makers missed the outrage of what China was doing. They were so wedded to the idea that free trade was the one true path to prosperity that they assumed the rise of China, a self-proclaimed Communist country, was the result of free-market economics.

If manufacturing jobs went to China, it was because of cheap labor. Or because China had a comparative advantage in manufacturing certain products. Strangely enough, those were the very same product categories that China had declared it would dominate in the Made in China 2025 state industrial plan created in 2015.

The corollary to that lax posture was that it made no sense to react to unfair trade policies with anything but even more free trade. If a country erected trade barriers, it was only mining its own harbors, hurting its own economy. Other countries should forgo even bothering to fret over whether trade relations were reciprocal. The most promiscuous of the free-trade lobby argued that the United States could eliminate all tariffs on all imports, unilaterally, and let all other countries continue to tariff US goods—and even that would benefit both the United States and the global economy as a whole.

It is a stubbornly naive view, even more so because of how widely it was held by smart, educated elites. China discovered that its major trading partners—especially those in the United States—were so wedded to a free-trade policy that they would stand aside and take the beating China was giving them. For years.

The free-trading nations were basically handing China the opportunity to prosper at their expense. China could almost be said to have had no choice in the matter. If that was the way the rest of the world was going to behave, it would have been irrational of China's leaders to scrap their successful practice of predatory mercantilism and follow fairer practices. Unfair behavior and theft were more than just usual tactics in China; they formed the basis of its prosperity.

China never would have reformed its behavior under the faint restraint of harsh words in diplomacy or small, temporary trade sanctions. So long as China's system worked because no serious retaliation resulted, changing it would be off the table.

Until President Trump came along. At the signing ceremony in the East Room, he had every right to savor the victory. He and China had been on a collision course for decades.

Donald Trump was bashing China for its trade transgressions long before he ran for president. It was part of his aversion to bad trade deals, generally. The threat of imposing tariffs as the "stick" had always been part of it, the carrot being the withdrawal of the stick. That resolute stance stemmed more from heartfelt sentiment than a calculated vote-getting strategy. Back in the 1980s and '90s, trade and tariffs were a loser as a political issue. There wasn't yet a crisis, and the US political process can stay mired in stasis and gridlock unless and until things come to a crisis.

The Republican establishment supported supposedly "free" trade, as did the US Chamber of Commerce. The Chamber represented the US-based multinationals that wanted free rein to ship jobs and production orders overseas, sell more goods and services to international customers, and bring cheaper foreign workers into the United States to work at lower pay, including thousands of tech engineers from Asia.

The liberal side, for its part, prized free trade and globalization as a way to lift up the developing world. Elsewhere, some antitrade lobbyists were said to be reluctant to ally with Trump, the real estate

developer, whom they regarded as a gossip column staple with sim-
plistic views. As a policy wonk in Washington, DC, Clyde Prestowitz
of the Economic Strategy Institute, told the *Wall Street Journal* at one
point, "You'd be tainted by his intellectually crude statements."

That quote appeared in a *Journal* profile of the president two years
after he won the election. It noted that he had "shifted many of his
positions" on the issues over the decades "and even changed parties.
On trade, however, he hasn't wavered an inch."

When Donald Trump ran for president in 2016, a lot of Americans
on the left dismissed it as the whim of a wealthy man looking for his
next gig a year after the end of his hit show on NBC, *The Apprentice*,
on the air since 2004. He was building his brand.

They underestimated him. His presidential aspirations had been
coming up for forty years. He had always believed that the United
States was getting cheated in free-trade deals, and he was certain that
one man could make all the difference. (Today we are required to add
"or woman," but that wasn't what he said at the time.)

He did his first TV interview in 1980, when he was thirty-four
years old. The host was Rona Barrett, a well-known gossip columnist trying her hand as a TV interviewer. The tape never aired, but
video of it exists on the internet, just like everything else nowadays.
He spoke in the same fragmented, repetitive style we hear from him
today, unrehearsed and unfazed by it. The video offers a prescient
exchange.

At the end of their chat, Barrett asked the young billionaire, "Why
wouldn't you dedicate yourself to public service?"

TRUMP: Because I think it's a very mean life. [Especially for]
somebody with strong views that are, maybe a little bit, which
may be right but may be unpopular. . . . The country, if we had
the one man, and it's really not that big a situation, you know,

people say "well, what could anybody do as president?" One man could turn this country around. The one proper president could turn this country around. I firmly believe that.

BARRETT: If you lost your fortune today what would you do tomorrow?

TRUMP: Maybe I'd run for president [snickered], I don't know.

End of interview.

Seven years later, in 1987, Donald Trump placed a full-page ad in the *New York Times* picturing a typewritten letter "To The American People," a cluttered, single-spaced missive of many words, signed at the bottom in that same bold, oversized signature he now used to sign bills, executive orders, and the new China trade deal. The letter began: "For decades, Japan and other nations have been taking advantage of the United States." The closer: "End our huge deficits, reduce our taxes, and let America's economy grow unencumbered by the cost of defending those who can easily afford to pay us for the defense of their freedom. Let's not let our great country be laughed at anymore." That ad ran more than thirty years ago, and it sounds as though he said it yesterday. Simply insert "China" in the place of "Japan." He says the two are interchangeable.

Then in 1999, as he was dropping hints that he might run for president, the *Wall Street Journal* published an op-ed "By Donald J. Trump, a real estate developer in New York." The headline was all Trump: "America Needs a President like Me." The lead ran:

Let's cut to the chase. Yes, I am considering a run for president. The reason has nothing to do with vanity, as some have suggested, nor do I wish to block other candidates. I will only run if I become convinced I can win, a decision I will make later this year. Two things are certain at this point, however: I believe nonpoliticians represent the wave of the future, and if elected I would make the kind of president America needs in the new millennium.

The eighth paragraph of the op-ed could have been tweeted by the president yesterday:

I also understand that our long-term interests require that we cut better deals with our world trading partners. This will raise an outcry because we've fallen into the habit of mistaking the easy availability of cheap, sweatshop-produced products for solid and sustainable economic stability. If President Trump does the negotiating, we'll get a better deal for American workers and their families, and our economy will not be as vulnerable to global pressures as it is today. Watch our trade deficit dwindle.

Over the years, the trade issue began to draw more attention among voters and politicians. In early 2011, Trump did an on-air phone call with CNN anchor Wolf Blitzer to say he was considering a run for president. He led with China.

TRUMP: They're making $300 billion a year and probably more than that each year—let's call it profit—off the United States.

They're manipulating their currency. Intellectual property rights and everything else are a joke over there. They're making stuff that you see being sold all the time on Fifth Avenue, copying various, you know, whether it's Chanel or whatever it may be, the brands, and just selling it ad—ad nauseam. I mean this is a country that is ripping off the United States like nobody other than OPEC has ever done before. . . .

These are not our friends. These are our enemies. These are not people that understand niceness.

Vintage Trump. He then outlined slapping a 25 percent tax on Chinese imports to raise hundreds of billions of dollars and added, "I would lower the taxes for people in this country and corporations in this country and let China and some of the other countries that are ripping us off and making hundreds of billions of dollars a year, let them pay."

He made that on-air call to CNN on January 20, 2011. Six years later to the day, Donald John Trump was inaugurated as the 45th president of the United States of America.

A year after the CNN interview, he endorsed the Republican nominee, Mitt Romney, who publicly praised Trump as "one of the few people who stood up and said, 'You know what? China has been cheating.'" Later, Trump would consider Romney for secretary of state and endorse him for US senator from Utah, which helped Romney win the election.

Then in the impeachment trial of President Trump in 2019, Romney would cast the lone Republican vote in favor of removing him from office. It was the first time in US history that a senator had voted to convict a president from his own party in an impeachment trial. Shameful. And ungracious.

By the 2016 presidential campaign, China was a more visible threat. The establishment elite was fighting harder against the Trump agenda. In late June, before the general election, the candidate made his first major visit to a swing state, Pennsylvania. He made a rousing speech in the city of Monessen in the western part of the state. A Republican hadn't won the state since George H. W. Bush in 1988.

The speech would prove to be a pivotal moment in the Trump run for the presidency. He had struck a deep chord in the blue-collar

crowd. As the candidate gave his speech in Pennsylvania, the US
Chamber of Commerce took potshots at him on Twitter, live-
tweeting ripostes: "Under Trump's trade plans, we would see higher
prices, fewer jobs, and a weaker economy." "Even under best case
scenario, Trump's tariffs would strip us of at least 3.5 million jobs."
Wrong, and wrong.

The next day, the *New York Times* led with this: "Donald J. Trump
vowed on Tuesday to rip up international trade deals and start an unrelenting offensive against Chinese economic practices." Which was
exactly what he later did.

A nice nugget in the *Times* story: "At nearly every campaign rally,
Mr. Trump has knocked trade deals with China as unfair to the American worker so frequently as to make his percussive pronunciation of
China a hallmark of impersonators." It would become Alec Baldwin's
stock in trade on *Saturday Night Live*.

The East Room signing ceremony was one of the most anticipated
political events in the three years of the Trump administration—and
in a lot longer than that. "Everyone" was present, as the president
would say, including some of the same people who had bitterly opposed the president and his tariffs from the start.

He made sure to recognize them publicly and thank them. It was a
gracious gesture for him to grant his foes coveted access to this historic event—that, or an opportunity to gloat tacitly right in front of
them. Either way, they had shown up.

Imagine listening to any other politician hold forth like this and
staying interested. President Trump pulled it off. At one point he told
the crowd:

So we have tremendous numbers of people here, and I'm saying, "Do I
introduce them?" But I think I sort of should because what the hell. This

is a big celebration. And, by the way, some of the congressmen may have a vote, and I don't—it's on the impeachment hoax. So if you want, you go out and vote. [Laughter.] It's not going to matter, because it's gone very well. [Laughter.] But I'd rather have you voting than sitting here listening to me introduce you, okay? [Laughter.]

At such a triumphant gathering, other presidents would have come across as solemn, august statesmen preparing to unveil a momentous milestone that would secure their place in history. And they would have hewed, obediently, to the script.

In the Trump Century, a diametrically opposed style has taken hold. President Trump shuns the stentorian oratory of his predecessors in favor of doing it his way: casual speech, off the cuff, laced with few rhetorical flourishes and abundant hyperbole, splashes of braggadocio, an eagerness to charm and impress, and unrelenting optimism, all of it delivered in a blunt New York style. You wanna piece of me?

Nobody President Trump knows is average or mediocre; everyone is "incredible," "tremendous," "brilliant," and more. Nothing he pursues is middling; it is the biggest ever, the fastest ever, the something-est ever, always. The president has no casual friends; in his speeches, all of them are his close friends, and in his eyes, they are the best ever at what they do. This strikes fans as amusing and touching. His haters see it as pathetic, insincere, and, perhaps worst of all to them, unpresidential. Horrors.

President Trump began the show by thanking a "very, very good friend of mine"—President Xi Jinping of China, his adversary like no other:

I want to thank President Xi, who is watching as we speak—and I'll be going over to China in the not-too-distant future to reciprocate—but I

want to thank President Xi, a very, very good friend of mine. We've—
we're representing different countries. He's representing China. I'm rep-
resenting the U.S. But we've developed an incredible relationship. But
I want to thank him for his cooperation and partnership throughout this
very complex process. Our negotiations were tough, honest, open, and
respectful—leading us to this really incredible breakthrough. Most peo-
ple thought this could never happen. It should have happened 25 years
ago, by the way. But that's okay.

Next, as the official transcript of the event records, the president introduced China's vice premier, who was nearby and would be making
a speech via an interpreter:

A man who also has become a good friend of mine and somebody who's
very, very talented and very capable, we're delighted to be joined by Vice
Premier Liu He, Ambassador Cui, and many other representatives from
the People's Republic of China.

That kind of patter rankled the hard-core hawks on China in the
United States, yet it is also what the Chinese like to hear. They desire
respect and credibility for their impressive rise up the economic ladder. They can be flattered, and the president is happy to do it if they do
what he wants them to do.

Then came a cascade of thanks and recognition to a procession of
players who had a stake in this happy occasion. The president recognized the great Henry Kissinger, who was secretary of state to President Richard Nixon when they opened US relations with China fifty
years ago. That changed the world. He was seated in the front row
directly next to me, on my left. He seemed to be enjoying it. Then I
heard the president say, "A man who always liked me—because he's
smart, so smart." He wisecracked, eliciting laughter, "The great Lou

Dobbs. You know, at first, he said, 'He's the best since Reagan.' Then, he got to know me more and more, and he said, 'He's even *better* than Reagan.'" Laughs all around. And across the United States, I am sure, Trump haters groaned at Trump's distinctive blend of mock bragging and obsequious praise of a cable TV host. It was gratifying and at the same time made me feel a little sheepish.

Then the president greeted the gentleman sitting next to me on my right, Michael Pillsbury, one of China's harshest and earliest critics. In 2006, a headline in the *Washington Monthly* called him "Panda Slugger." He has advised President Trump and three Republican predecessors. Now the president was thanking him, welcoming his wife, and saying "And you've been saying some fantastic things about China and about us. And we have a good partnership. This is going to be something that's going to be very special. We're going to talk about it in a second."

He was holding back the news a little longer. And then it occurred to me: my seating placement had been purposeful. It must have been by design. I was seated between the two polar ends of the China argument: openness and accommodation versus guardedness and confrontation.

Minutes later, President Trump pulled off another perfect pairing, two people present who were diametrically opposed on the Trump tariffs. First, he introduced his longtime backer Dan DiMicco, an ex-CEO of the steel producer Nucor Corporation, a veteran in the struggle to compete with China's supercheap, unfairly subsidized steel. He had been an advisor to the Trump campaign, and now the president told him, in front of all of us, "Dan DiMicco. We've been fighting the steel things for 20 years together, Dan. I'd like to say 30, but I don't want to do that. Right? We've been fighting together for a lot longer than 20 years on the—"

"We're winning," DiMicco shot back.

"And now we're winning," the president said, taking back the baton. "Finally. It took 25, 30 years. It took a little change at the top, didn't it? But you've been a warrior for—for getting really taken advantage of as a country. And you've been a warrior, and I appreciate it. Right from the beginning, Dan and I—we—it was a fight of two people against the world."

The very next name he mentioned was that of a steadfast opponent of the Trump tariffs: the CEO of the Chamber of Commerce. "Tom Donohue, US Chamber of Commerce. And that's—where's Tom? Thank you, Tom. Great job." That was gracious of the president, given the group's unrestrained carping at him in the past.

Eventually the president turned back to the script, dipping in and out to ad-lib a few more accolades. The phase one deal he and the Chinese vice premier had signed that day called for, as the president put it at the ceremony, "substantial and enforceable commitments regarding the protection of American ideas, trade, secrets, patents, and trademarks."

And restrictions on currency devaluation. The president told the room, with the vice premier of China nearby:

> And, in all due respect, China was one of the greats in history at doing that [currency valuation], and we're going to work on it together. But currency devaluation will now have some very, very strong restrictions and some very powerful restrictions. . . . One of the strongest things we have: total and full enforceability.

He repeated that phrase, as he is known to do. Jimmy Two Times. I have heard other politicians start doing this more since his arrival. The president then cited China's obligations to thwart counterfeits and stop forcing US companies to give away their trade secrets to Chinese partners they were forced to wed.

"So now," the president told the CEOs in the East Room, "when Boeing has some work done over in China or wants to sell planes over in China, they don't have to give up every single thing that they've ever—you know, that they've worked so hard to—to develop and to come up with. Are you guys hearing that? You don't have to give up anything anymore. Just be strong. Just be strong. Don't let it happen. But you don't have to do that."

He was on a roll: "It was a terrible—it was a terrible situation going on there. And a lot of it was because our co—our companies, I have to say this, were very weak. You were very weak. You gave up things that you didn't have to give up. But now, legally, you don't have to give them up."

The president went on to explain why the Trump tariffs would stay in place despite China's sizable givebacks. The president explaining, with disarming transparency that might have unnerved Henry Kissinger, that he left them in place as a chit for negotiating further concessions in phase two:

We're leaving tariffs on, which people are shocked, but it's great. But I will agree to take those tariffs off, if we are able to do phase two. In other words, we're negotiating with the tariffs. We have 25 percent on $250 billion worth of goods. And then we're bringing the 10 percent down to 7.5 on $300 billion worth of goods.

So—but I'm leaving them on, because otherwise we have no cards to negotiate with. And negotiating with Liu is very tough. But they will all come off as soon as we finish phase two. And that would be something that some people on Wall Street will love, but from what I see, they love this deal the way it is now. But we have very strong cards. And, frankly, China and I are going to start negotiating with Bob and Steve and everybody very, very shortly.

And then all of us in the East Room could tell that President Trump had returned to his script: "So, from this nation's vibrant heartland to our gleaming cities, millions of workers and farmers and innovators have waited decades for this day." So had Donald Trump.

He rattled off features of the new trade pact: The Chinese had agreed to spend $200 billion on US goods in the next two years. That would include $50 billion on US farm foodstuffs, $75 billion on US manufacturing: "They'll be putting into our country, okay? They're going to be putting into our country $75 billion on manufacturing." Plus $40 billion to $50 billion on US financial services, including MasterCard, and $50 billion worth of energy.

The president: "So that's great for our energy people. We're the number one in the world now; we weren't. We're now the number-one energy group in the world. We're bigger than Saudi Arabia, and we're bigger than Russia. We're bigger than everybody."

The president's math was optimistic: he said $200 billion and had just itemized a buying list totaling $225 billion. He likes to think big. Bigger. It was another exaggeration, though, given the festive air of anticipation and celebration, it went unchallenged at the ceremony. No doubt the fact-checkers at CNN were scurrying after proof that the president had lied again.

His detractors say he makes things all about himself, but the salesman in him makes them all about the people he knows. He ran through them in a certain succession. His first toss was to the vice president, Mike Pence. Then the US trade representative, Robert Lighthizer, who had gone after the Chinese on intellectual property and negotiated theft restrictions in the new accord. Then Treasury secretary Steven Mnuchin, who had opposed the tariffs at first. ("And thank you very much, Steve. Great job.")

Then his son-in-law, Jared Kushner, and his daughter Ivanka

Trump. The president knows this drives the libs to distraction. Nepotism!

The president further celebrated the presence of icons of big business, who had typically opposed the Trump tariffs but might now benefit very well. Early on in the ceremony, he started by calling out "a friend of mine," Steve Schwarzman of BlackRock Financial, with more than $7 trillion in assets under management. It holds licenses to sell funds to institutions and high-net-worth investors in China, where the asset management industry totals $14 trillion. BlackRock looks to Asia to drive 50 percent of the growth in industry assets in the next five years, with China as the driver.

The president quipped: "Steve, I know you have no interest in this deal at all." (Applause.) He recognized UPS chief David Abney ("Thank you, David. Great job you've done."). And ConocoPhillips CEO Ryan Lance: "Ryan, great. You're doing fantastically well. Most of you, I can say, you're doing fantastically well. 'Thank you, Mr. President.' [Laughter.] Don't worry about it. Don't feel guilty."

He even managed to take a swipe at the Fed and its exorbitantly high interest rates, upon spotting former Federal Reserve governor Kevin Warsh in the audience. Trump had considered him for Fed chairman before naming Jerome Powell. Considering how much the president had been bashing the Fed and Chairman Powell during the past year, on Twitter and elsewhere, Warsh may have been relieved at this outcome. The president: "Kevin Warsh. Kevin. Where's Kevin? I don't know, Kevin. I could have used you a little bit here. Why weren't you more forceful when you wanted that job? Why weren't you more forceful, Kevin? You're a forceful person. In fact, I thought you were too forceful, maybe, for the job. And I would have been very happy with you."

He mentioned that other countries with riskier balance sheets are charging investors to borrow their money, while the United States is still paying investors interest: "I love this. This concept is incredible. Again, you don't know where the hell it leads. But you borrow money, and when you have to pay it back, they pay you. This is one that I like very much. And I'm going to talk to you about that, Lou Dobbs."

There it was. In his meandering style, the president had put more pressure on the Fed chairman and finished with a callback reference to me, a cable anchor who might take up his cause. Man, this guy is *good*.

The China win will become a centerpiece of the Trump Century. It ends decades of neglect and abuse in trade and reins in the United States' most threatening nemesis. The new deal stops short of ending our trade hostilities with China. Those hostilities will continue for years to come. That is less the point.

President Ronald Reagan said he was willing to negotiate with the "evil empire," the Soviet Union, so long as he could "trust but verify." Meaning: he didn't trust the Russians. The presidents who follow Trump will have to avoid trusting the Chinese. However, they will have the armor of the hard-and-fast restrictions the Chinese just agreed to follow.

The president went back onto the prompter, mostly, and his words rang every bit as true as when he was winging it with alacrity:

As a candidate for President, I vowed strong action. It's probably the biggest reason why I ran for President, because I saw it for so many years. And I said, "How come nobody is doing something about it?" In the meantime, immigration, and building our military—also important. But that's probably the biggest reason.

In June of 2016, in the great state of Pennsylvania, I promised that I would use every lawful presidential power to protect Americans from unfair trade and unfair trade practices. Unlike those who came before me, I kept my promise. They didn't promise too hard but— [*applause*]—*they didn't do anything. And I actually think I more than kept my promise. . . .*

This is an unbelievable deal for the United States. And, ultimately, it's a great deal for both countries. And it's going to also lead to even a more stable peace throughout the world.

CHAPTER SEVEN

JOBS, JOBS, JOBS

P resident Trump's thirty-year obsession with unfair trade was linked inextricably to three other strategic goals that drive his presidency: jobs, jobs, jobs.

That focus helped him pull off his underdog victory in the 2016 election, and it was the reason he won the swing states Michigan, Ohio, Wisconsin, and especially Pennsylvania, where he had visited a recycling plant that June and made the big speech that launched his bid to topple decades of free-trade orthodoxy in both parties.

The president's passion for job creation owes less to a lust for votes and more to an abiding desire to make America grow again. He is driven by the belief that stronger GDP growth in the United States will create better job growth, which can lift wages where it is needed most, at the low end of the pay scale and in blue-collar factory jobs in what our trade failures led us to call the Rust Belt of the Midwest.

Every president hopes that a rising tide will lift all boats. President Trump was the first to say he wanted the tide to reach the battered boats that hadn't been afloat in a long time. In the first three years of Trump, the president had racked up a startlingly successful record on job growth. And wages. So many people said it was impossible to do. We did fabulously well, and the signs were that things would get even better, right up until early 2020 and the Wuhan crisis.

The crisis wiped out all of the progress we had made and cut far deeper than that in terms of both jobs and the stock market. President Trump wants to get it all back. As destructive as this setback has been for the country, we can recover fully from the economic losses, faster than ever before, and grow from there.

It will take a lot longer for us to heal from the emotional toll of losing so many Americans to the coronavirus contagion. With time, growth can help us on that front, too.

Some people will argue that the Wuhan pandemic and the damage it did to commerce and people's livelihoods are far worse than the Great Depression. In the virus crisis, upward of 25 million Americans filed for jobless benefits in the first two months. The nation's GDP had its steepest decline in our history, falling by an estimated 20 to 30 percent, perhaps even more.

The Great Depression lasted a decade; thousands of employers went out of business, and the jobs they provided were lost forevermore. The Wuhan catastrophe erupted when Americans were enjoying one of the strongest economies in our history. At the time, President Trump continually called it *the* strongest economy ever. His predecessors had given up trying to rebuild the US manufacturing base and told us the economy would have to move into services, where pay at the low end is far lower than in the factory.

In the aftermath of Wuhan, government forced businesses to lay off tens of millions, making far more people dependent on government for their livelihoods, at the risk of breeding more dependence, indolence, and binge watching of Netflix. The recovery benefits from the government, approved in the immediate crisis by lawmakers looking to "do" something, paid out more to some people than they had been earning in their regular jobs. Why go back to work until you absolutely must?

Not his way. President Trump and Congress threw trillions of dollars at the Wuhan recovery, but he would prefer that the United States grow its way out of problems and crises. We can do this by freeing business and investment to help us rebuild faster than any New Deal can achieve. Business must run better to survive; government has an almost endless supply of taxpayer dollars regardless of how incompetently it makes use of them.

President Trump knows it in his bones: we can grow our way out of this dystopia. It requires our taking the right steps to unshackle the economy, cultivate business restarts, and get people back to work. This is the president's strong suit. An examination of the miracles he pulled off in job creation in the opening three years of his term, before Wuhan, offers hope for what is possible.

In the Trump jobs boom, all groups of the population benefited, and blacks, Hispanics, women, and the lowest rung of earners fared even better than the population overall. It felt so good at the time. We were *back*!

Employers created 6.807 million new jobs in the three first full years of the Trump presidency. Every single month posted jobs growth, continuing a positive streak that had been ongoing and unbroken since October 2010. Real weekly earnings—how much higher pay rose than the inflation rate—grew at a strong annual rate of 2.5 percent.

Job openings, another sign of growth, stood at almost 7 million in late 2019, a rise of 21 percent in three years. The unemployment rate fell to 3.5 percent, from 4.6 percent when President Obama had left office, tying the lowest level set in 1969. The jobless rate leading into

2020 had been less than 4 percent for twenty-two months in a row, the best streak in fifty years. The median historical jobless rate since 1948 has been 5.6 percent.

This happy upturn was a function of a humming economy and strong corporate profits, which rose at an annual rate of 5.6 percent in President Trump's first three years. That fueled a rise of 47 percent in the stock market, based on the S&P 500 index.

The job growth, in turn, improved the lives of people at the bottom of the income ladder. The poverty rate fell almost a full percentage point to 11.8 percent. Some 6.3 million Americans were able to get off the food stamp rolls because they no longer needed the extra help. That marked a 15 percent decline in benefit recipients to 36.4 million, the lowest number since late 2009.

The promising jump in jobs was so startling that the policy makers at the Federal Reserve started to fret even more than usual about the prospect of rising inflation. The better the jobs numbers, the higher the risk that the Fed would raise interest rates again, which would crimp the growth that President Trump was working so hard to deliver. That was why he started hammering so hard on his appointee as Fed chairman to stop raising rates and start lowering them. (See chapter 12.)

The growth in jobs came despite the Democrats' attempts to oust President Trump from office, which creates uncertainty, which then rattles business and investors. The upturn occurred despite mean-spirited criticism in the left-wing corporate media outlets and opposition from corporate giants and leading lights in his own party. And in spite of trade jousting with China.

Imagine how much hotter job growth might have been had the tormentors of the president dropped their pitchforks and torches and joined the Trump revolution.

The most remarkable aspect of the surge in new jobs lies in the sector hit worst by years of unfair trade: manufacturing. It had been left for dead by everyone else, who said that the jobs we had lost never were coming back. But President Trump was bent on rebuilding the base. These are the blue-collar, get-your-hands-dirty jobs that once paid solid wages, even to those who never attended college. That was the heart of what built the middle class from the 1950s onward. Until China came along and spoiled everything.

It seems an unassailable goal. Stoke growth in the economy to fund expansion, and start pressing US companies to bring home most of their supply lines. Yet the Democrats, the media, the elitist establishment, and other denizens of conventional wisdom were resolutely set against it—against even trying.

Settling for less is alien to Donald Trump. His upbeat attitude and unbridled spirit—and even his bent for overpromising and exaggerating—are vital assets. Especially in the monumental effort to rebuild the US economy if he won reelection. This is one reason the Fake News media has counted thousands and thousands of lies told by President Trump; he has a gift for hyperbole.

On President Trump's watch, manufacturing jobs started coming back for the first time in years. In the first twelve months of the Trump reign, manufacturers added 207,000 manufacturing jobs. In year two, we added 284,000 jobs. Through March 2020, the number of blue-collar jobs was up by 491,000 since Trump's inauguration.

Impressively, the once shrinking manufacturing sector added jobs at a faster pace than the rest of the economy in the first year. By March 2020, durable goods positions—good factory jobs making things intended to last for three years or longer—exceeded 8 million for the first time since the Great Recession. The category had added 350,000 new jobs since Donald Trump had become president.

Even in the sectors hit most directly by the Trump tariffs, job growth surprisingly rose. The numbers rolled in despite predictions of disaster. After the steel tariffs went into place, jobs in metals-using manufacturing actually *rose*. As the year 2019 opened, durable goods manufacturing—the biggest user of metals—posted more than three hundred thousand job openings, almost a 17 percent gain from before the tariffs.

Far from depressing employment in the metals sector of the US economy, the tariffs had been accompanied by a boom. President Trump's most vehement critics should be able to celebrate these numbers, but they were unmoved. Can't they ever be happy for him? (That was a joke.)

When the Trump Tariffs took aim at steel imports in March 2018, the doubters had a new objection. Previously they had argued that he would fail to create the jobs he had pledged to create. Now that jobs growth was improving, they said his new regime of tariffs would destroy the new jobs they had insisted were impossible to produce. He could not win.

The Dems, actually, are even more cynical than that. Before President Trump took office, they started painting his push for a renewal in US manufacturing as driven by a hidden agenda: it was racist. They floated the notion that any renaissance was driven by President Trump's desire to help whites rather than all Americans.

The fear and loathing went beyond the op-ed pages to infect supposedly straight news. The economics correspondent for the *Washington Post* described the Trump victory as "The Revenge of Working-Class Whites." On CNN, a network analyst said that Trump's economic promises were "the rhetoric of white nostalgia."

In other words, in the minds of the president-elect's critics, even before he took office, his push for a manufacturing revival somehow was a call for white supremacy. No wonder they wanted so badly for it to fail.

It is worth pondering the underlying source of this widespread aversion to efforts to rebuild the United States' manufacturing might. It had been proven wrong so quickly that it must be based on something other than sound economics and reasoned analysis. Rather, this antipathy was rooted in a loathing of the United States' manufacturing legacy, for the strong blue-collar country of our not-so-distant past. Or a loathing of its people.

Or, more likely, a loathing of President Trump.

A factory with a short supply of qualified workers, happily, is open to hiring all comers. In fact, a 2016 report put together by unions and manufacturers explored how hard the decline in factory jobs had hit African Americans. Trump understood that good jobs are for everyone. The notion of "white nostalgia" would be obliterated by reality in the next three years, anyway. The Trump economy was good for every US demographic segment, especially blacks.

In Trump's first three years, black employment rose to an all-time high of more than 9 million people. The black jobless rate hit an all-time low (6.6 percent) by early 2020, as did the jobless rate for Hispanics, which fell to 5.4 percent.

Also, in the first three Trump years, the wide gap between black and white unemployment fell to the narrowest ever recorded. For decades the black jobless rate had been stubbornly stuck at about twice that of whites, both when the economy boomed and when it slumped. That began to change under President Trump.

By the end of his second year in office, black unemployment stood at 6.6 percent versus 3.4 percent for whites. The black rate had dropped to 1.85 times the white jobless rate, rather than double (2.0). By August 2019, the gap had narrowed so that black unemployment was less than 1.6 times white unemployment. It amounted to a 20 percent reduction in just the first three years of the Trump administration.

That really is some "white revenge," huh? What self-respecting racist president would let that happen?

• • •

By March 2018, as President Trump was slapping steel tariffs on China, the United States reached a milestone unseen in more than a hundred years. Job growth was so strong that unfilled openings outstripped the total ranks of the unemployed. At one point, we had 6 million people unemployed and more than 7 million vacant jobs.

In November 2019, the economy added 261,000 hires, 53 percent higher than the average monthly job growth for the previous year. Everyone can be happy with that, right?

The Democrats remained unshaken in their pessimism. House Speaker Nancy Pelosi bemoaned the good news: "Despite some encouraging numbers, the November jobs report offers little solace to the farmers and hard-working families who are struggling to stay above water with the costs of living rising and uncertainty surging."

What uncertainty? Farmers make up 1.3 percent of the US workforce. Inflation was tame. For working America, in fact, average hourly wages were up 3.1 percent, more than a point above the rate of price inflation. Before he was elected, the Democrats doubted that candidate Trump could raise US employment, and now that he was president, they refused to acknowledge that he had. The left wing's all-out hatred of President Trump meant that even good news was bad news in his first term. They must have wondered: How could the Trump wizardry possibly have worked?

We tend to forget how bleak the future looked before the dawn of the Trump Century. Cut to the decade leading up to 2016 and the presidential campaign—eight years after the start of the Great Meltdown and after eight years of Obama. A phrase I always disliked, "the new normal," was the new buzzword for our slack job creation. As if to say, that's just the way things are. The Dems would resurrect it in attempting to suppress the comeback push by President Trump after the Wuhan lockdown ended.

President Obama had run for the White House on the slogan "Hope." By his exit, economically, at least, we had lost a lot of it.

When he ran for office, candidate Trump declared audacious goals that strained the foundations of economic credulity: 25 million new jobs in the next ten years, up to 4 percent goal for GDP growth, faster-rising wages, yet with low inflation. Once again, the doubters and haters said he never would pull it off. Truth is, he would fall short of those ambitious targets because he always aims high.

Even nonpartisan experts at think tanks and at the Federal Reserve said it was undoable. They insisted that the US economy was capable of tepid, milquetoast growth at best. Long-term unemployment would stay mired at 5 percent no matter what. GDP growth would be stuck at 2 percent if we were fortunate. The United States' aging population would fail to supply enough new workers. Live with it.

From the start, candidate Donald Trump promised a brighter vision. His can-do bravado and sky-high promises rankled many observers. President Obama took it personally. In June 2016, with five months to go before the 2016 vote for Hillary or Trump, Obama traveled to the blue-collar city of Elkhart, Indiana, the site of his first visit as president eight years before. There he delivered a speech that made it sound as if he were running for a third term.

He criticized the Trump contention that international trade and immigration have hurt American voters and labeled as "crazy" the candidate's proposal to roll back regulations on Wall Street to stoke economic growth. He preached to the crowd in Elkhart, "That will not help us win. . . . That is not going to make your lives better." It is an interesting locution he deployed often: always about the "not."

Later that day, President Obama did a town hall meeting on PBS in which he hammered the Republicans' likely nominee without deigning to mention him by name on national television. As if that somehow might hold back Donald Trump. The cable networks had bypassed the Obama speech in Elkhart earlier that same day, using bites rather than running it live. The speeches by candidate Trump, though, got live national coverage in full, even on the networks that despised him. Trump drew better ratings. That had to irritate President Obama a little. He was the president, after all.

The PBS town hall produced a sound bite that went viral, and years later, the printed text of it captures the Obama speaking style perfectly. You can hear him saying it even now:

> *When somebody says, like the person you just mentioned who I'm not going to advertise for, that he's going to bring all these jobs back, well, how exactly are you going to do that? . . .*
>
> *He just says, "Well, I'm going to negotiate a better deal." Well, how—what—how exactly are you going to negotiate that? What magic wand do you have?*

Some candidates might have softened their proposals and gone vague after getting a public dressing down from the president they hoped to succeed. Candidate Trump grew even bolder. Two months before election day 2016, he spoke before the Economics Club of New York at the New York Hilton hotel. A teleprompter foul-up forced him to ad-lib at the opening ("It's a good thing I brought notes"), and the candidate began by bragging about how well he was doing in the latest polls.

The teleprompter rebooted, and he unveiled his plan for an "American economic revival," a regimen of tax cuts, deregulation, and better trade deals that shredded all current thinking. Though he had posted his plan on his campaign website, it was the first time he was presenting it publicly, and he had picked an audience of influential executives and Wall Street money managers for that moment:

> *It's time to start thinking big once again.*
> *That's why I believe it is time to establish a national goal of reaching 4% economic growth. . . .*
> *Over the next ten years, our economic team estimates that under our plan the economy will average 3.5% growth and create a total of 25 million new jobs.*

It was an outrageous claim and instantly provoked shock and disbelief. Trump's declaration struck many as crazy. "Trump's 25 Million New Jobs Promise Doesn't Add Up," CNBC declared. The mainstream media and economists had decided that the US economy was incapable of growing that quickly, that it was already near maximum employment. Manufacturing was bound to continue its long, irreversible decline.

At the time, the idea of reviving US manufacturing and creating jobs in the tattered sector was such a long shot that the elites viewed Trump voters as being deluded. At the Brookings Institution, a center-left wonk tank, a senior fellow, Mark Muro, flatly decreed, "Trump won't be able to 'make America great again' by bringing back production jobs."

Even people who agreed that "free" trade had hurt US job growth said his was an impossible crusade. On the political website FiveThirtyEight, an economics writer, Ben Casselman, posted an article that captured the defeatist mind-set before the 2016 vote for president. To argue the contrary was unsophisticated or unintelligent.

The article, "Manufacturing Jobs Are Never Coming Back,"

began, "A plea to presidential candidates: Stop talking about bringing manufacturing jobs back from China. In fact, talk a lot less about manufacturing, period." So why even try? The article continued:

> It's understandable that voters are angry about trade. The U.S. has lost more than 4.5 million manufacturing jobs since NAFTA took effect in 1994. . . . [T]here's mounting evidence that U.S. trade policy, particularly with China, has caused lasting harm to many American workers. But rather than play to that anger, candidates ought to be talking about ways to ensure that the service sector can fill manufacturing's former role as a provider of dependable, decent-paying jobs. . . .
>
> Whether or not those manufacturing jobs could have been saved, they aren't coming back, at least not most of them.

The cause of bringing back manufacturing jobs was hopeless. A responsible politician would focus on managing the ongoing decline of US manufacturing and building more jobs in the service sector—including, say, more jobs serving customers at McDonald's, though that chain has been installing ever more automated, ATM-like kiosks to let customers order without human contact.

After President Trump won the 2016 election, Casselman doubled down on the doom: "The larger problem for Trump and his supporters is that there is very little reason to think that any set of policies could meaningfully reverse the long-term decline in U.S. manufacturing jobs."

A few weeks later, on November 30, the Wharton School, one of the most respected business schools in the country, posted an article online quoting its professors on what they expected to happen next. An emeritus professor, Stephen J. Kobrin, asked melodramatically, "What happens when people realize they've been taken? When people realize that he can't bring back jobs and that they are not better off than they were two years ago, how does he use it—who does he blame it on?"

President Trump hadn't even moved into the Oval Office, and already assumptions were that he would fail, that he was lying the whole time, and that he will do something ominous and threatening, soon. What the hell is wrong with these people?

The anti-Trump views were held with such conviction by such a wide array of Respectable People. The punditocracy never would have made such a sad sack, feckless forecast about the policies of another newly elected president. Certainly not the first black president, Barack Obama. Or the first woman president, if Hillary Clinton had won.

In some ways, this was just more Trump hate. President Trump succeeds in spite of it—or maybe even because of it. An old saying posits that you never should underestimate the value of having a great adversary. President Trump revels in the fight. He is energized by the force of the opposition. It makes him stronger, like some superhero who is able to absorb all the negative energy coming at him and convert it into more power for himself and his agenda.

Set ambitious targets, reach further than anyone else, and you can end up advancing much further than you otherwise would. It is a most Trumpian trait; it is the promoter in him. Let the lib media say he lied by being too optimistic. The Dems will say he broke a promise. President Trump always aims too high.

Now Trump's promise of a stronger US economy no longer seems crazy. The unemployment rate fell to 3.5 percent in November 2019, the third anniversary of Trump's election. A month later, the Federal Reserve held a policy meeting, and the chairman told reporters afterward that the underlying rate of unemployment now seemed to be much lower than anyone had expected.

Anyone except Donald Trump.

• • •

Only one sector of the US economy had been able to spare its workers from having to struggle with the many ravages of global trade and technology disruption: government workers at all levels.

Many years ago, working for the government meant accepting a below-market salary to be in public service. It was a sacrifice. Today a government job pays significantly better than business, both in pay and in benefits—a combined premium of 17 percent, based on a CBO study released in October 2017.

The federal leviathan employs 2.2 million workers in six hundred different occupations, spread across more than one hundred agencies. This amounts to 1.34 percent of the entire US workforce. It spends $215 billion a year on compensation. That works out to a per worker cost of $97,000 annually. The government has 32 percent more staff than the largest US employer, Walmart, with 1.5 million employees.

In the federal government, employees with only a high school diploma earn the highest premium compared with the same workers in private companies: 53 percent more. Compensation is 21 percent higher for people with a bachelor's degree. Those premiums expanded robustly in the Obama years; ten or fifteen years ago the premium for government work was 20 percent higher than in the private sector. This surge occurred in spite of a freeze on across-the-board raises from 2011 to 2013.

The soaring costs of benefits for government workers, especially pensions, are the biggest driver of the surge in federal pay in recent years. For the least educated federal worker with a 53 percent edge over private sector pay, the wage premium was 5 percent—and the value of related benefits was 93 percent higher. College grad employees earned 5 percent more in salary and 21 percent more in benefits than did people in the business world.

Government workers are blissfully unaware of the hardships besieging their counterparts in private industry. They face no threat from offshoring and being fired so that their jobs can move to cheaper foreign markets. They face less competition from lesser paid outside service providers. The feds also have stringent rules against hiring immigrants who are in this country illegally. Any immigrant employees come from the population who already live in the United States legally. No H-1Bs for federal jobs.

The government can pamper its workers with great benefits and high salaries and cover it by taxing and borrowing all the money needed. The private sector has to earn it by selling something of value. Public employees are also unaffected by the creative destruction of technology. It has wiped out several million jobs and created plenty of others, while government worker rolls expand year after year.

One reason so few federal workers support President Trump is that none of them were struggling like their private sector counterparts before his election. Nor have they benefited from the economic resurgence he has ignited. When the leader of our country made a public appearance and waved to the crowd at a home game for the Washington Nationals in the 2019 World Series, he was roundly booed. There are people who live in DC and don't work for the Deep State or the Swamp, but I haven't met many.

Compensation for federal workers is unaffected by the unfair trade that hurts private workers. Civilian federal employees working in the Defense Department, the largest group, face no wage competition from defense workers in China or Pakistan. That is true for almost all federal workers. From the Department of Agriculture and the Department of the Interior to the US Treasury and the Justice Department, the wage competition faced by employees is purely domestic.

Wouldn't that be nice?

Even some Republicans offer the wrong explanation for this spread between government and private pay. They say the feds are paid so well because the government has an unlimited budget.

That is an invitation to socialism. Tell blue-collar Americans that only the government can afford to pay decent wages, and they will be more likely to support a government that keeps growing and employing more people. The idea becomes more attractive that all jobs should be government jobs; all benefits should be government benefits. No wonder "Medicare for all" has gained support.

Wages in the private sector are lower because employers have withheld sharing the surging profits they have reaped. Corporate profits soared once the economy recovered from the financial crisis. Companies raced to buy back their shares because profits were so excessive they lacked a better use for the returns. Cash was piling up on corporate America's balance sheets faster than it could be spent.

Was the problem greed? That is what Senators Bernie Sanders and Elizabeth Warren would tell you. The wages in the private sector were low, they would say, because American corporations were greedy, while the US government was benevolent. That, of course, was a humblebrag: they were praising themselves. It missed the reality that corporations have always been greedy and the people in government no less so.

Government workers have another edge their counterparts lack in business: powerful and obstinate unions. The little-recognized fact is that labor unions represent less than 7 percent of private sector workers in the United States and some 40 percent of government workers at the federal, state, and local levels.

Big labor has had a seat at the table in Washington for decades. AFL-CIO president Richard Trumka visited the Obama White House many times. Unions did great things decades ago, when few federal laws protected workers, and helped set standards for working conditions. Today the labor movement is largely a government worker movement.

This lends itself to a self-perpetuating symbiosis. Unions contribute more than one hundred million dollars directly to political campaigns each election cycle; by some estimates, their combined spending on all political activities eclipses $1.7 billion. They help elect those who will protect their pay and increase their pensions, and the government's long-term obligations for pension payments have soared as a result.

The higher pay for public employees feeds higher dues collected by the American Federation of State, County, and Municipal Employees (AFSCME), the Service Employees International Union (SEIU), and other government workers' unions. They can then spend more on political donations to the pols who keep approving further increases. Nobody has tried to stop this.

The slanted mainstream media never reported the real successes of President Trump on the jobs front. They held on to their grudge, and they were too busy chasing their own tails, distracted by whatever was the next outrage of the next day, whether it was the president's usage of a banned word ("He called it 'Wuhan virus'!") or some other offense. Maybe he does this to them on purpose. It distracts them from covering the real things he is accomplishing, and this way, they have less chance to complain about it.

For all the job growth, it helps the economy less if wages fail to grow, as well. For twenty or thirty years, price inflation often has risen faster than wages. This was due to big, tectonic-plate kinds of problems that were viewed as being almost impossible to solve.

Abundant obstacles exist to unlocking growth in compensation. Sluggish GDP growth and years of shipping jobs overseas were only the start. More people were going to college later in their twenties. Digital disruption of millions of jobs in dozens of industries was another factor. In its wake came a mismatch of skills and needs between those who want jobs and those who want to hire.

As well, almost 40 percent of able-bodied adults were out of the workforce, even before Wuhan. President Trump has lamented this trend. The labor participation rate fell by almost 3 percentage points during the Obama era. By December 2017, it finished up 0.4 percentage points, at 63.2 percent of the workforce.

In the first term of President Trump, pushing up wages would be harder to do than ever before. Although he had an answer for that, too: higher GDP growth can spark businesses' hunt for new investments that can help create new jobs. Once again the experts scoffed; they said wage growth was stuck for sophisticated reasons in these interconnected times of global trade.

Once again, President Trump would prove the doubters to be wrong.

CHAPTER EIGHT

HIGHER WAGES

No matter what the outcome of the 2020 elections, the Republican Party never will be the same. Its new direction and outlook will last decades after Donald Trump leaves the stage. The party's longtime predilection for big-business interests and crony capitalism has yielded to a focus on the working class in the heartland.

From fat cats to Joe Sixpack. From pinstripes to denim and flannel.

He has forged the Republican Party into a party that pursues the interests of working-class and middle-class Americans. His election demonstrated the folly of the Republican establishment's plot to win the presidency through promoting the agenda of its corporate donors.

Trump carried Pennsylvania, Ohio, Michigan, and Wisconsin by adding working-class voters to the party's base. He made working-class Americans the center of his politics, and he took the entire Republican Party with him. He consistently draws above a 90 percent approval rating from party voters. Up until the Wuhan virus lockdown, his rallies drew capacity crowds with thousands of extra fans gathering outside the venues. Even today, some party leaders resent this refocus.

American workers have been rewarded for their support of Donald Trump. After years of falling or stagnant wage growth, wages began a sustained rise in late 2017. In September 2017, average hourly earnings growth surpassed 2.8 percent for the first time since the Great Recession a decade ago. Since August 2018, average hourly earnings have grown at a rate of 3 percent or more, month in and month out.

Wage gains for "production and nonsupervisory employees"—the American working class—hit the 3 percent mark in August 2018 and quickly rose above 3.5 percent. In October 2019, workers were earning on average 3.8 percent more than they had the prior year. Inflation was at 1.76 percent. So at that point, workers' wages, after chronically failing to keep up with inflation for decades past, were growing at *double* the inflation rate.

Impressive. And rarely, if ever, did you hear any celebration of it by the left-wing media.

Better still, the best gains were going to workers at the low end of the pay ladder. Americans without a high school diploma experienced wage growth of *nearly 6.5 percent* in 2018, the largest hike among all educational demographic groups. This was a reversal of the Obama era when wage growth was concentrated among workers with the highest incomes. Before the Dems spent quite so much time wailing about wealth inequality.

American workers were winning again.

When wages limped along in the Obama era, employers were too stingy to pay more to their workers. They were guilty of this long before he took office. After the 2009 financial crisis, wage growth failed to rebound as the US economy tiptoed into recovery. Wage increases were subpar for years afterward.

From 2010 through 2016, the last year of the Obama administration, wage growth for Americans managed to hit 2.5 percent only once, in 2016, and fell as low as 1.5 percent in 2012. For the first three years of Obama's second term, average wage growth ran 2.16 percent per year—compared with a 2.96 percent rise, on average, in the first three years of Trump. That is 37 percent higher growth for the Trump tenure.

Bravo! As my Twitter followers heard from me on June 19, 2020, the president had created 3.1 percent real wage growth at the time. That would rise to 3.5 percent later, a nice jump. More Americans were working than ever before, and he seemed a sure bet to be reelected because of it. Dems' messaging, meanwhile, hewed to their standard socialism, in stark contrast with Trump's buoyant form of capitalism.

It remains to be seen if workers can keep on winning. With wages barely rising for ten or twenty years, productivity—how much in goods we produce for the same unit of work—rose handsomely at a higher rate. The upside went to the owners of capital rather than to the workers.

Rather than celebrate the reawakening of wages, the elite and the resistance to President Trump saw a darker picture. Corporate America and the nation's central bankers believed that something must be done to stop those wage gains. They still feared the specter of inflation, looming but nonexistent for years. The better Trumponomics started looking, the more prone the Fed was to raising interest rates.

This is why the president also was waging a relentless campaign to press Federal Reserve chairman Jerome Powell—his own pick to run the central bank—to stop raising rates and start cutting them deeply. (See chapter 12.)

America's elites had become so comfortable with secular stagnation, and so spoiled by cheaper labor overseas and low wage growth in the United States, that they greeted even the slightest upward pressure on wages with panic.

The global elites at the nation's huge trading companies, financial institutions, multinationals, and tech titans benefited from globalization, cost cutting, layoffs, offshoring, and stagnant wages (with rising productivity, of course). As wage growth started waking up in the Trump era, they soon concocted a cause for this alarming trend: the United States suffers from a severe labor shortage. That must have been why wages had been rising handsomely since Trump took office, especially at the low end—up almost 7 percent, more than triple the pace of inflation.

Some authorities, starting at the conclusion and finding a way to get there, said that the United States is incapable of providing enough workers from its own population, which is why we must open our gates and let millions of immigrants rush into our country to fill service jobs—albeit the service economy was supposed to be the crutch that would let us limp along after losing our manufacturing base.

Experts cite myriad reasons for this purported labor shortage: tepid long-term GDP growth in the United States; the aging population as ten thousand people turn age sixty-five every day; our low birth rate. This leads to intellectual arguments and more studies, this time on why the United States is in for a future of stagnant wages and workers can skip seeking higher pay altogether. This creates more self-defeatism and reasons why we are barely able to make it out of bed.

It was as if the elites had been gaslighting Grandma to make her think she is more feeble and frail than she really was.

This argument by the globalist elites is a laughable canard. Even before the coronavirus, almost 40 percent of able-bodied Americans under age sixty-five were out of the workforce and sitting at home. It is a shocking figure. The labor participation rate, as a percentage of all people who could be working, is near 62 percent and has been in decline for many years. This shows how much richer the United States is than we realize. More important, it is proof that many Americans were unmotivated by the stubbornly low pay that companies continued to offer.

If too many eligible workers stay out of the workforce, the solution is simple if we believe in a truly free market: businesses must pay better starting salaries to persuade the idle to reenter the workforce. But both large and small employers often can avoid paying more for low-end workers because millions of undocumented immigrants in the United States illegally will work for below-market wages. Thus the focus on fomenting talk about the labor shortage.

This imaginary labor shortage is a favorite reason for explaining why, to hear the Democrats tell it, the United States is doomed to become another Japan, on life support and in recession for thirty years—unless we import more people from overseas. Lack of workers was one of the biggest concerns that businesses reported to the Federal Reserve System for its "beige book," a collection of anecdotal evidence about economic conditions gleaned from conversations with owners in each of the Fed's twelve districts. The phrase "labor shortage" became a mantra repeated endlessly in the mainstream left-wing media.

In May 2017, the *New York Times* published an article with the startling headline "Lack of Workers, Not Work, Weighs on the Nation's Economy." The story included a number of anecdotes about alleged worker shortages—most of them showing nothing more than the fact that wages were rising—which the media usually hailed as a *good* thing.

One construction company claimed it had raised its starting wage by 10 percent to almost $18 an hour and still was unable to attract sufficient workers. The free-market answer to that problem is to raise wages by 15 percent or to whatever level will persuade workers to quit their jobs and come work for your business.

The data show just how little tolerance business leaders have for wage growth. The average weekly earnings of construction workers in Utah stood at $970 in April 2017, according to the Bureau of Labor Statistics. That was a rise of just $30 a week from a year earlier, up 3.25 percent. Weekly pay had been *higher* four years prior to that, at $980. The numbers are not adjusted for inflation, so the actual decline has been steeper.

But somehow, offering a 10 percent raise and failing to find an abundant pool of grateful, unemployed workers was evidence of a labor shortage—at least, it was evidence to the *Times*. The truth is that companies in Utah and across the United States were slow to raise wages even as unemployment fell to record lows. They were spoiled by having spent so many years having their pick among the oversupply of millions of unemployed people.

Workers, for their part, had spent so long toiling amid low wage growth and knowing they were imminently replaceable that they were unaccustomed to asking for raises. That is their fault. Even today, the rate of workers voluntarily quitting their jobs to get something better is very low by historical standards. The jobless rate fell to an all-time nadir in the first three years of President Trump's first term, and the quit rate should have been higher.

It was as if the tool of offering higher wages to attract new workers had fallen out of the favor among US businesses and no one had noticed because it hadn't been necessary to do so for so long. But now that far fewer unemployed people were desperate for any job, employers profess to be unable to find staff. It is as if they have forgotten how to poach workers from competitors, or perhaps they remain too stingy to start paying up—and are in tacit collusion, across the land, to keep things that way.

One of the sources of low wage growth, even as unemployment declined, may have been companies adding less experienced workers and part-time consultants instead of expanding their full-time staff. These newcomers tend to earn lesser wages, and their hiring reduces the need for full-timers. Their presence can make experienced workers hesitant to demand better pay.

But as the Trump boom took hold and pushed our economic expansion into the longest on record, companies found it harder to find new workers—at the lower-than-market rates they wanted to pay, at least. More baby boomers were hitting retirement age, forcing companies to compete for younger workers to restock their ranks.

Another upward force on wages had to be the surging optimism that had taken hold in the American people and the US economy following the election of Donald Trump. Consumer sentiment soared and then remained at historic highs month after month. Consumers are also workers, and that confidence may have led them to seek higher wages for the first time in years. Give me a raise, they say, or I may hunt for a job elsewhere.

What's more, workers knew that Trump had staked his political fortunes on their economic fate. He had taken on powerful actors in the economy that many workers see as opposed to their interests, from chief executives to China. When General Motors set plans to close a plant in Ohio, Trump went after it on Twitter, brushing off attempts by management to blame the unions. Even before he took office, he had shamed Carrier into scuttling plans to close a furnace factory in Indianapolis and move production to Mexico.

Even that small victory was under scrutiny by the *New York Times*. A reporter revisited the Carrier site almost two years later, publishing a gainsaying story on August 10, 2018. The headline positively jeered, "At Carrier, the Factory Trump Saved, Morale Is Through the Floor."

The article admitted that local unemployment in Indianapolis was down to a low 3.3 percent, the factory had saved 700 jobs, and it had recalled another 150 workers who had been laid off. The plant had the capacity to turn out eleven hundred new furnaces a day. The problem, the *Times* said, was worker absenteeism:

> *What's ailing Carrier isn't weak demand. . . . Instead, employees share a looming sense that a factory shutdown is inevitable—that Carrier has merely postponed the closing until a more politically opportune moment.*

It added:

> *In some ways, the situation is a metaphor for blue-collar work and life in the United States today. Paychecks are a tad fatter and the economic picture has brightened slightly, but no one feels particularly secure or hopeful.*

The better things started to look for higher wages, the worse the supposed labor shortage was reported to be, in the media, at least. On April 3, 2018, a website in Monterey Bay, California, reported that restaurants were struggling to find workers: "Qualified cooks and dishwashers have all but disappeared, and in response local restaurants have lowered their standards while raising pay. Current applicants have weaker skills . . . and cooks with just a few years experience are applying for jobs better suited to those with a decade or more on the line." The piece quoted an executive chef as saying "We pay dishwashers $14 an hour ($3 above minimum wage) to offset (the shortage). . . . To be honest, it's the most integral position. If you can't put food on a clean plate you have nothing to sell." The article noted that in the past, "restaurant owners could always keep kitchen salaries in check because candidates were lining up around the block."

After years of being at a disadvantage in negotiating with employers,

US workers were gaining sway and had more choices. More power for the workers—usually, this would be something the media would hail heartily. When it happens under President Trump, the response is muted at best. That's okay, honey badger don't care, as the funny YouTube video put it some years ago; it applies well to President Trump.

Two weeks after the Monterey Bay story ran, the media meme pushing panic over a supposed labor shortage culminated in a piece in the *Wall Street Journal*. It ran in the paper's news pages.

The *Journal*'s op-ed pages long have been an outlet for advocating open borders and low wages. Run by longtime insiders in Washington, the *Journal* editorial page carps at the president more than it praises him. Star columnist Peggy Noonan, long ago a speechwriter for President Reagan, tacitly despises him. Karl Rove, the Republican consultant who helped elect George W. Bush, publishes smart columns that offer Democrats advice on how to run against him. Even from the most conservative paper in the United States, President Trump gets less support and praise than he has earned.

The news story continued the new agenda: "The U.S. is facing a severe worker shortage, forcing employers big and small to explore the labor market's youngest echelon, which is piling into the workforce." Average wage growth for production and nonsupervisory employees that month had been just 2.8 percent. Yet the paper claimed that businesses were experiencing "historic, severe worker shortages." Well, they should pay more, so that wages rise at, say, a 3.5 percent or 5 percent rate, and then check on whether the fake labor shortage persists.

The article contained a chart showing that the employment rate of American teenagers had been rising, as if to support its premise. Zoom out to a wider view, and it turned out that teenage employment remained far below prerecession levels. In fact, fewer teenagers than ever were seeking employment.

It had escaped even the *Journal* that that was less evidence of a "very dire" labor shortage than, perhaps, the fact that wages were still too low to interest an American teenager in taking a low-level job.

No one knows for sure why teenagers stopped working. But the timing of it suggests two factors: trade and immigration. The steep decline began a few years after the North American Free Trade Agreement was signed in 1994 and accelerated after China was fully admitted to the World Trade Organization in December 2001.

Just as trade with China and Mexico under NAFTA had hit certain US communities especially hard, it had reduced teenage employment in particular. At the same time, the foreign-born share of the US population was hitting a historic high. Guest workers and immigrants, both legal and illegal, could be hired to do the jobs that American teenagers had once filled.

The effects of this are writ large in the summer resort business in the nation's posh playgrounds, including Martha's Vineyard in Massachusetts, preferred by President Obama, and East Hampton on Long Island, a beach retreat for media titans and Wall Street bigs. Another hotspot is Nantucket Island, a Quaker outpost off the coast of Massachusetts that was converted long ago into a summer resort for America's wealthy. It provides a case study of what happened to jobs for American teenagers.

Nantucket was a prized summer job location for the young men and women of New England a few decades ago, within living memory of some of the longer-term residents of the island—the ones who were there before the hedge fund managers and lobbyists sent the prices of summer cottages to unimaginable heights.

The youngsters arrived at the island for summer break from high school or college to work as golf caddies, dishwashers, waitresses, and groundskeepers—the kinds of jobs filled by immigrants today. Now Nantucket, like many seasonal resort areas across the country, is entirely dependent on foreign workers for the low-end jobs teenagers once filled, such as bagging groceries.

Hiring teenagers in summer jobs at the resorts had been on the wane since 1990. The same year, a foreign guest-worker program opened in the United States and began letting in a new influx of service staff by the thousands. More young people sought internships and other college-oriented pursuits rather than a summer of cash from manual labor.

In June 2005, the *Times* celebrated the legions of "fresh-faced students from Bulgaria, Poland, and Lithuania." But a "labor shortage" created by delays in the guest-worker program had put a crimp into summer plans for the hoi polloi:

> But this year, a guest-worker shortage could cripple the season there, and at many other resorts. In Colorado, the Broadmoor Hotel in Colorado Springs was denied all 250 visas it customarily receives from the federal government for housekeepers, landscapers, and masseurs. In Michigan, the Yankee Rebel Tavern on Mackinac Island is trying to make do without its usual staff of 18 dishwashers from Jamaica. And in Florida, Amelia Island Plantation has no one to help manicure its golf links.

The paper also lamented, "The shortage also means fewer foreign faces. The guest-worker program, created in 1990, had the unintended effect of transforming formerly apple-pie resorts into virtual Epcot Centers of languages and cultures." The shortage was so severe that the *Times* included this somber note: For the second year in a row, the labor shortage could mean long waits at restaurants, shorter menus, untidy hotel rooms, reduced store hours, and poor service.

The playgrounds of America's elites had become inhospitable to young Americans. The foreign workforces that dominated their payrolls were openly hostile to American summer workers, whom they viewed as taking "their" jobs. Often an individual restaurant or business would be staffed entirely by workers from one country. The *Times* article reported that nearly all thirty-five bus drivers employed by the Nantucket Regional Transit Authority were foreigners, mostly from Bulgaria. Being able to speak the foreign language or the local dialect of the workers became more or less a job requirement. So a kid from South Boston had no chance of making it as a dishwasher, bus driver, or baker.

Once upon a time in America, teenagers complained about having to work too much. Sixty years ago, Eddie Cochran lamented in the song "Summertime Blues," "Every time I call my baby, try to get a date, my boss says, 'no dice son, you got to work late.'" But by 2016, most American teenagers were idle, and they shunned even trying to get a job. Work was for foreigners.

This is a lamentable change in our economy, mostly unnoticed in our society. Traditional American values have a foundation in our ability to improve our lives through hard work and determination. The crash in teenage jobs that began in the 1990s was a warning sign of what was happening across the United States. It received scant attention.

Most Americans think of themselves as members of the middle class, whether they are young men and women toiling at a digital start-up, earning minimum wage at a part-time job, assembling trucks in the Midwest, or practicing law at a white-shoe law firm. Throughout our history, we were the Land of Opportunity, rather than the Land of Envy that we hear a lot of politicians talking about now.

We admired the achievements and financial rise of those who earned it, without idolizing or envying them. We respected those below us on the income ladder and their efforts to climb higher. We saw ourselves as able to rise above any restraints put on us by dint of our lineage or the lack of it. We believed that hard work gave us the opportunity to improve our circumstances for ourselves and our children and grand-children.

The higher wage growth ignited by the Trump agenda came without any government-ordered hike in private wages. Democrats have long pushed raising the minimum wage as a sop to voters and a great claim to be helping the poor. Let us order parsimonious employers to pay more or be in violation of the law, they say. The old Soviet Union took a similar approach to wage growth before it collapsed economically and politically.

The Dems wanted to force companies to pay a higher minimum wage of $15 an hour nationwide—whether a person worked in New York, where prices are extortionate, or in rural Idaho, where prices are lower.

This is a bad idea. If raising the minimum wage were the path to prosperity, if it really worked, the congressmen of both parties would have lifted starting pay to $100 per hour by now. Just force evil big business to pay for it.

A higher, legally mandated minimum would do nothing to help workers or lift their wages. The United States had upward of 155 million people employed full-time in the boom times before the Wuhan pandemic. Fewer than 1 million of them were in minimum-wage positions. Typically they stay there a year and move up the pay ladder—without government intervention.

Raising the minimum pay to $15 an hour from the $7 or so of recent years would force employers to double the pay for entry-level jobs. It would prompt them to fire upward of a million of their lowest-paid workers and use more part-time staff, according to the nonpartisan Congressional Budget Office. This would deprive even more American teenagers of a shot at early entry into the first jobs of their lives.

Many states mandate higher minimum wages than the feds do, and that is their right. New York forces businesses to pay almost $12 an hour, whether the employee is a burger flipper or an insurance adjustor. The states know best, at a local level, what works for them. Still, letting the free market determine where rates should go is a smarter way to operate—if employers will pay fairly and play by the right rules.

To the extent that raising the minimum wage forces employers to bump up the pay of more experienced workers the next few levels up, it means it was done by government fiat rather than by market forces. Ultimately this kind of thing helps crush thriving economies. It is a masked form of socialism if taken too far.

After President Trump had racked up strong growth in wages, his lib enemies had to find another way to throw shade at it. Weeks after holding his one-man show in the East Room to sign the China trade deal, he made a triumphant State of the Union address on February 4, brimming with energy and can-do possibility.

Four days later, an "exclusive" article appeared in *USA Today*, citing the real reason for the impressive growth under President Trump: it had been caused by *states* that had raised the minimum wage. What? The headline was "Trump Touted Low-Wage Worker Pay Gains but Much of the Credit Goes to State Minimum Wage Hikes."

The beginning of the story then attempted to undercut any credit President Trump can claim, noting that he had called attention to a blue-collar jobs boom and rapidly rising wages for low-income workers: "He's correct, but Trump left out one thing: a large portion of those gains can be traced to minimum wage increases in more than half the states."

The national newspaper cited a study "provided exclusively to USA TODAY by the National Employment Law Project (NELP)," saying that median wages for the bottom one-fifth of workers had "climbed much more sharply in states that have raised their pay floors than in states that haven't." The story argued that the president's program had made no difference at all in the long run: "More broadly, real pay for the bottom fifth of workers nationally increased 3.6% from 2009 to 2018 while wages for all other workers were stagnant, the NELP figures show."

That story got huge attention in the lib media. There had been no progress at all. How could that be? The answer lies—and I do mean lies—in intentional deception or abject depression.

The *USA Today* article failed to mention that the report supporting minimum wage hikes came from a group that promotes minimum wage hikes. NELP is a 501(3)c nonprofit advocacy group. Its website lists its achievements: "Won $15 minimum wages in NJ, IL, CT, and MD. Won local unemployment compensation legislation in D.C. for furloughed workers during the government shutdown."

NELP makes no claim to being nonpartisan. Its fund-raising is run by ActBlue Charities, which assists Democrats. Its board includes Jared Bernstein, a former Obama advisor who was chief economist for Joe Biden; Sharon Block, a former deputy assistant secretary in the Obama Department of Labor; Elyssa McBride, secretary-treasurer of the American Federation of State, County, and Municipal Employees (AFSCME); labor lawyers; and assorted NGO executives. None of that was revealed in the story.

The paper cited "more than half of states" (twenty-six), but really it was less than half (twenty-three). The study included states that had small wage bumps for cost of living that were too tiny to have helped; they should have been excluded. The report framed a five-year period starting in 2013 for states that raised pay but used a nine-year period starting in 2009 to measure how much median wages rose. This is known as data dredging.

The last paragraph of the *USA Today* story said that in early 2020, two dozen states and forty-eight cities and counties would raise the minimum wage in their markets. Cities and counties at $15 an hour would double to 32 localities. The story neglected to say that it was a sign of economic strength in the era of Trump. State and local pols would avoid doing so if their economies were too feeble to withstand it.

The paper also reported that even the nonpartisan government Office of Management and Budget (OMB) had issued a report to Congress in late 2019 saying that the Trump administration's "efforts to reduce taxes, eliminate regulations and implement fair trade deals are driving economic growth and increasing workers' take-home pay far more effectively and efficiently than legislation." The report was a response to a Democratic bill to raise minimum pay nationwide to $15 an hour.

Trump wins again.

Never again will either political party ignore the working people in jeans and flannel shirts in Flyover America. That will be another lasting ramification of the Trump Century. Other candidates of all stripes were too sophisticated and worldly to decry the decline of wage growth in the United States and the loss of US jobs to cheap labor in factories in Asia. CEOs, boards of directors, and investors were too focused on maximizing shareholder value and their own upside.

President Trump came to the rescue, and the Dems and the media still were in denial about it. He had confronted China in the crisis over the Wuhan virus. And thrown down the gauntlet of tariffs on a quarter-*trillion* dollars' worth of Chinese imports to the United States. Rather than triggering a trade disaster, it got China to the table to make concessions and sign a new deal that will benefit the United States. Yes, huge.

That helped US wages head upward at higher rates than we had seen in years. The new China accord and the tariffs would contribute further to that. None of it would work, however, without putting into place another critical piece of President Trump's five-pronged strategy: immigration reform.

It would become one of the most bitterly fought issues of the Trump presidency. He saw it as vital to protecting national security and reviving job growth: protect American jobs, and avoid undercutting our wage growth with cheap, imported labor; ensure strong borders and controlled immigration to let in those who can help our country most and refugees in special cases; block violent criminals and terrorists from slipping past our defenses.

It would have benefited the nation to have a debate on those grounds. The president's opponents preferred to debate immigration policy on just one ground: President Trump, they said, was a racist.

CHAPTER NINE

IMMIGRATION AND JOBS

As the Wuhan pandemic carved a destructive pathway across the United States in early 2020, it created clear proof that President Trump had been right about so many things: we must tighten our control of immigration, get tough on China, harden our borders to stop illegal entries, and assert the right to shut out any persons we see as a danger or an enemy and stop them from moving to our country.

Yet the crazy and obstructionist Never Trumpers lambasted every move the president made to get out in front of the Wuhan emergency and cut off the entry of potentially infected foreigners who were trying to enter the United States.

On January 31, President Trump issued an executive order prohibiting entry into the United States for all foreign nationals who had visited China within the previous two weeks. On February 3, he cited security concerns and added Sudan, Tanzania, Eritrea, and Nigeria, all in Africa, Myanmar in Southeast Asia, and Kyrgyzstan in Central Asia.

Trump foes saw the new order as another attack on civil liberties, immigrants, fill in the blank. The ranks of the woke instantly reviled him for it, especially on Twitter. On February 3, the *New York Times* published an op-ed by Jamelle Bouie, a staff columnist who had joined the paper in 2019, with the headline "The Racism at the Heart of Trump's 'Travel Ban': Adding Nigeria to the expanded list of excluded countries just makes it more obvious."

Then, on April 20, the president issued a new order temporarily banning *all* immigration while the government scrambled to contain the coronavirus and its aftermath. Cue same old, same old. Democrats raced to grab a piece of the tweet stream and were almost foaming at the mouth.

Hakeem Jeffries of New York, the House Democratic Caucus chairman, called Trump "Xenophobe. In. Chief." Congressman Joaquin Castro of Texas called the new order an authoritarian-like move to take advantage of a crisis. Congressman Bill Pascrell of New Jersey called it a "poisonous distraction" to divert blame for the administration's failures. Kamala Harris, a senator from California and failed candidate for the Democratic presidential nomination, said Trump was politicizing the pandemic to double down on his supposedly anti-immigrant agenda.

By then, for President Trump, that was all in a day's work.

We used to speak of a politician's "playing the race card." Today for the Democrats, the race card amounts to the entire deck: *everything* is racist in some way, especially when it regards President Trump and the Republicans. From the start, the president's immigration policy has been based on common sense as part of an overall strategy to create more jobs, spur higher wage growth, and ensure the security and safety of the American people.

The premise of the Trump doctrine in immigration is simple, clear, and defiantly self-centered: America first. We cannot afford to take in all of the millions of impoverished, unskilled refugees who would come to our shores if they could, fleeing socialism, dictatorships, political repression, terrorism, drug cartels, and natural disasters. We must pick and choose. Who gets in should be based on who will best serve our needs.

Polls show that more than 150 million foreigners would move to the United States if they could. Once you agree that it is a bad idea to let in *everyone* who would like to come here, you have found common ground with Donald Trump. Now it is a matter of debating the limits and deciding where to set them.

It reminds me of the well-known story about Winston Churchill's asking a matronly woman if she would ever sleep with a man who agreed to pay her a million dollars. When she said yes, he asked whether she would accept ten dollars. When she got offended, he is said to have told her that they had already agreed she was for sale; now it was a matter of price. Churchill never actually said it, but the joke got a lot more leg out of the rumor that he had.

First, some big numbers: the foreign-born population in the United States quadrupled in forty years to 40 million immigrants from 1970 to 2010. An additional 4.4 million have arrived in the ensuing seven years, according to the Pew Research Center. That is a rate of almost 630,000 new arrivals annually. This increased the percentage of foreign-born residents from 13 percent to 16 percent of the US population.

In 2015 an estimated 12 million illegal aliens resided in America, up half a million in a year, based on Department of Homeland Security figures. Roughly 1.7 million had been here less than five years, while 80 percent (9.6 million) had been here longer than ten years and 6 percent for five to ten years.

From 2000 to 2007, almost half a million illegals resettled in the United States every year; then the economic crisis intervened. In the years 2010 to 2015, the rate dropped to 70,000 per year. Apprehensions

at the US-Mexico border in 2017, the first year of the Trump administration, totaled 415,517. They jumped by 25 percent the next year to
eclipse half a million—and then soared by 87 percent in 2019.

A key reason US wage growth was so stunted for so long, especially in entry-level jobs, was the ample supply of immigrant workers
residing in the United States both legally and illegally. Illegals, especially, are willing to work under the table for below-market cash pay,
which is tax free for both parties.

Various economic studies say that for every 10 percent rise in the
migrant population in the United States, US wages fall by an average
of 2 percent. Some people argue that it makes no difference. Others,
such as the Harvard University economist George Borjas, maintain
that a 10 percent increase in immigration could cause up to a 4 percent
reduction. The Borjas numbers would be cited by the Trump administration in support of its aims.

The president's opponents refuse to believe that his intent on immigration is related to jobs and wage growth. They see only one thing,
every time: a racist.

It began from the moment he announced his run for the presidency
on June 16, 2015, which feels like a hundred years ago. He and his
wife, Melania, a former fashion model, glided down the golden escalator of Trump Tower and into the ornate marble-and-brass lobby
packed with press and cheering supporters. Some people were writing
off the idea as a publicity stunt in his search for an encore to *The Apprentice*, which had ended its eleven-year run on NBC a few months
earlier.

Trump announced his candidacy in front of a raucous crowd in a
blunt, unvarnished lexicon that was foreign to US politics. Out were
the clichés of the past, the lofty language concealing underhanded
schemes to boost donors' fortunes. Out, too, were the phrases carefully sculpted by insights from focus groups to appeal to special

interests while offending no one. In a frank conversation confronting the economic crisis of our country in our time, he put immigration and trade front and center.

Instead of delivering a sterling, rehearsed pronouncement and hewing to the script, the new candidate winged some of it:

> *Our country is in serious trouble. We don't have victories anymore. We used to have victories, but we don't have them. When was the last time anybody saw us beating, let's say, China in a trade deal? They kill us. . . .*
>
> *When do we beat Mexico at the border? They're laughing at us, at our stupidity. And now they are beating us economically. They are not our friend, believe me. But they're killing us economically.*
>
> *The U.S. has become a dumping ground for everybody else's problems.*

The lobby of Trump Tower crackled with applause and cheers. Just a minute or so into the speech, Trump had already demonstrated that he was a very different kind of candidate. He was a break from the past. He was going to stand up for the American people. But even as the onlookers cheered, they had no idea what was coming next. He continued:

> *When Mexico sends its people, they're not sending their best. They're not sending you. They're not sending you. They're sending people that have lots of problems, and they're bringing those problems with us. They're bringing drugs. They're bringing crime. They're rapists. And some, I assume, are good people.*

He then declared that US political and corporate leaders had failed. They were "losers" and "people who don't have it":

We have people who are morally corrupt. We have people that are selling this country down the drain.

That was the first time in a generation that Americans heard a candidate for president speak directly and without fear or hesitation. To many people present that day, he sounded like a free man should sound, almost a figure out of the mythical past of colonists, pioneers, and cowboys in the Wild West. He was running because there was no one else who could get the job done. His language, shocking to liberals, was no big deal to his supporters; he talked like they did, and some of them thought it was funny.

The media and political elites of both parties failed to see it. They had grown too accustomed to looking for gaffes. They had long ago abandoned reporting what candidates said, or planned to do, in favor of covering anything off message or mistaken.

They declared that Trump had called all Mexicans rapists, though he hadn't. Anyone following news about Mexico had read about gang members and thugs who cross into the United States illegally, such as those loyal to MS-13, a notorious Salvadoran immigrant gang formed in Los Angeles and with 10,000 members in LA, New York, Boston, and other cities. More than 100,000 illegal immigrants are in US prisons, as well as more than 50,000 legal immigrants—and almost 2 million native-born Americans.

From the *New York Times* to CNN to the *Washington Post*, the left-wing corporate media assured themselves and their audience that Trump's presidential bid was dead on arrival, having committed suicide by political incorrectness. New York City's *Daily News* cover line was "Clown Runs for Prez."

It would become a familiar pattern. Trump would say something brutally frank or just a little too vague, and it would scandalize the radical Dems, the mainstream media, and even the GOP establishment.

They were becoming indistinguishable from one another in their fight against all things Trump. Eventually they melded into the irrational and destructive resistance that has persisted throughout the Trump years. With each new offense, they would declare that his campaign—and later his presidency—was finished.

But Trump's refusal to back down struck many other Americans as a healthy rejection of the timidity and rank cowardliness they had grown accustomed to sensing in political leaders. Politicians, particularly Republicans, were hindered by taboos against strong opinions stated in plain language. President Trump was unbowed and unafraid to offend. He was positively gleeful when attacking the sacred objects of the establishment of his own party, the media, and the radical Left.

The liberal media portrayed President Trump as appealing to the inner darkness of America. That itself is an insult to the American people, and it presumes they had hatred in their hearts that needed to be overcome by voting for Democrats. That may have been a hangover from President Obama, who after a terror attack or a police confrontation would warn us against having a knee-jerk racist reaction.

The growing number of Trump supporters saw something else: an outsider who spoke freely and didn't care if his words melted the snowflakes who were taking offense. The totalitarian Left had been training Americans to accept and demand that our politicians tread carefully and bow to anyone who took exception to almost anything: the wrong pronoun, the wrong Halloween costume, the wrong expression of faith.

President Trump refused to play this game. He pushed aside the provocations and prevarications of the libs to pursue redeeming US politics and restoring our prosperity, freedom, and unity. He realized that political discourse in the United States had broken down so totally that the country was failing to address fundamental challenges. The strictures against frank debate had left us unable even to recognize the problems we faced. In a media world where everyone demands apologies for everything, Trump apologized for nothing.

As soon as he took office, President Trump erred on the side of taking decisive action. Seven days into his first term, on January 27, 2017, he set off bitter outrage with a simple executive order on immigration. It was temporary, set to last ninety days. The reverberations and recriminations of it would extend for a lot longer.

President Trump signed an order banning entry by citizens from seven Muslim-majority countries: Iraq, Iran, Syria, Libya, Somalia, Sudan, and Yemen. Ninety days. That's it. The lib media and the Left instantaneously branded it a Muslim travel ban. Fifty nations around the world have Muslim-majority populations, says the Pew Research Center. Not seven.

The move sparked protests at airports in New York, Chicago, Los Angeles, Seattle, San Francisco, and more cities, as well as protests overseas. Two thousand people marched on John F. Kennedy International Airport in New York, snarling traffic, disrupting airport operations, and thwarting the travel plans of thousands of Americans. A few days later, 1,200 employees of Comcast, the owner of the liberal MSNBC and CNBC networks, walked off the job in Washington, Philadelphia, Sunnyvale, and Portland, Oregon. A few days more, and there were street marches in London, Paris, New York, and Washington.

On January 30, former President Obama put out a statement, though he had been out of office only ten days and historically US presidents have avoided second-guessing their successors. It said that he "fundamentally disagrees with the notion of discriminating against individuals because of their faith or religion." So does everyone, but President Obama had just implied that President Trump was doing just that.

Obama was revered in the Fake News liberal outlets for his

moralizing stance. Yet he was, in reality, firing a pious and sanctimonious jab at the man who had just destroyed Obama's plans for handing off his legacy to another Democrat, Hillary Clinton. He was trying to further damage Donald Trump, just weeks after instructing FBI director James Comey, in a joint meeting with other officials, to keep him apprised of developments in the Russiagate investigation of President-elect Trump.

What a fraud and a liar President Obama had turned out to be, hailed as so noble and yet driven by such dark, mean-spirited, unconstitutional tendencies.

That was a lot of anguished outcry over an order that was set to last three months and cover a handful of countries. The new administration wanted a pause to investigate the vetting process for admitting visitors and immigrants to have a way of identifying risks. For the protests, many of them orchestrated by NGOs and community groups, as always, the real purpose was to vent outrage at the election of Donald Trump.

It was the beginning of the Left's new strategy of conflating attacks on the failing or corrupt leadership of a country with the citizens of that country. Now if you criticize any country in Latin America, Africa, Asia, or the Middle East, it must be because you are racist. Express outrage at China's feckless efforts to contain the Wuhan virus, and you are anti-Asian.

Let's do a fast-forward zip through the inevitable court battle, as it shows the classic cycle of Trump action, followed by outraged reaction, back to Trump action, ping-ponging back and forth:

- US refugee assistance NGOs (nongovernment organizations) and Arab American groups, backed by dozens of leftist friends of the court, sued to block the new ban. They represented two Yemini brothers whose travel plans had been disrupted. They cited the president's anti-Muslim tweets and said that his blatant bias against Muslims made the order illegal and unconstitutional.

• Seven days after President Trump acted, a district court in Seattle issued a ban on the ban, Judge James Robart ruling. The media made the judge a folk hero for standing up to Trump. The *Washington Post* gushed, "Judge James L. Robart wore a bow tie to the hearing, opened with a joke and finished with a thunderclap. He was known for that sort of thing. . . . At the end of the hearing, with no jokes or spare words, Robart halted Trump's ban and potentially changed the fate of citizens of seven majority-Muslim countries and tens of thousands of refugees, who had been denied entry into the United States." Changed the fate of? That, over a ninety-day travel ban.

• President Trump was undaunted, tweeting at 8:12 a.m. the next morning, "The opinion of this so-called judge, which essentially takes law-enforcement away from our country, is ridiculous and will be overturned!" How dare he? the protesters demanded.

• A month later, in March 2017, President Trump unveiled a revised travel ban—and was again blocked, this time by an appeals court in Richmond, Virginia, one of two times it ruled against the Trump order. The ruling read, "Plaintiffs offer undisputed evidence that the President of the United States has openly and often expressed his desire to ban those of Islamic faith from entering the United States." It overlooked the key question: Did President Trump have the constitutional authority to issue the order?

• In June, the Supreme Court allowed part of the ban to take effect. In September, the Trump team put out a revised version of the order, taking aim at travelers from Chad, Iran, Libya, North Korea, Somalia, Syria, Venezuela, and Yemen. Two months later, in December, the Supreme Court let the latest version of the order take effect pending appeal. It was the first time the high court had let any full version of the travel ban go forward in its entirety. There were only two dissenters: Ruth Bader Ginsburg (a Bill Clinton appointee) and Sonia Sotomayor (an Obama appointee).

• And then, six months later, came another Trump triumph: on June 26, 2018, the Supreme Court upheld the president's constitutional authority to impose the temporary travel ban to protect the people of the United States. SCOTUS flatly dismissed the depiction of the executive order as a ban on Muslims, as the Left had insisted it was, saying "The text says nothing about religion."

In the ruling, split 5–4 with all five Republican appointees in favor of Trump and all four Democratic appointees voting against, Chief Justice John Roberts wrote the opinion for the majority: "The Proclamation [of the travel ban] is expressly premised on legitimate purposes: preventing entry of nationals who cannot be vetted and inducing other nations to improve their practices." In other words, it was "well within executive authority."

President Trump. Right again.

It was a satisfying vindication for the administration, albeit Mr. Chief Justice added a dig, noting that the justices "express no view on the soundness of the policy." Who asked him?

Justice Sotomayor wrote a vociferous dissent, joined by Justice Ginsburg. She attacked the court majority as much as the Trump ban, writing that the ruling was "all the more troubling" because of supposed parallels to *Korematsu v. United States*. That was the scandalous ruling the high court made during World War II upholding Franklin D. Roosevelt's authority to order the forced relocation and resettlement of 120,000 Japanese Americans on the West Coast after the Japanese attacked Pearl Harbor.

Translation: in a way, Justice Sotomayor had just implied that what President Trump wanted to do to Muslims now was akin to what FDR had done to the Japanese back then. It was ridiculous, and the libs loved it. The other two liberals on SCOTUS declined to join that overheated dissent, and Chief Justice Roberts took exception to it, noting that it was "wholly inapt to liken that morally repugnant order" in the World War II case to what had happened in this one.

All along, the resistance and the far Left had said that the president had violated the Constitution and was guilty of racist and fascist acts in issuing the temporary travel ban. All along, the president and his supporters had said this fell within his executive authority. He had argued that the United States should stop entries, temporarily, from a handful of nations racked by strife and terrorism until it could develop a more secure vetting process.

Now the Supreme Court of the United States had ruled that President Trump was right. A torrent of criticism erupted. Democratic National Committee chairman Tom Perez said, "Discrimination is not a national security strategy, and prejudice is not patriotism. Let's call this ban for what it is: an outright attack on the Muslim community that violates our nation's commitment to liberty and justice for all."

Even one of the lead attorneys arguing against the Trump travel order, Neal Katyal, refused to bow to the high court's reasoning, saying "We continue to believe, as do four dissenting justices, that the travel ban is unconstitutional, unprecedented, unnecessary, and un-American." The majority had ruled otherwise. Had conservatives reacted so brazenly defiant after a Supreme Court ruling, the media would have decried it.

President Trump had prevailed once more. In addition to strengthening national security at the borders, the other angle to his immigration policy had to do with protecting jobs and wages from the ravages of unfair competition.

It is a safe bet that every single person who tells you there is a labor shortage is a supporter of higher levels of immigration and lower levels of border security. The complaints are disguised demands for the US government to provide more cheap workers by importing them into our country or by stepping back from tariffs on imports from China. There is no US labor shortage. (See chapter 8.)

The official unemployment rate was very low, but it counts only Americans who were actively looking for work. More than 90 million able-bodied Americans lack a job and aren't looking for one. Even when you take into account retirees and college students, workforce participation is low. Just 82.6 percent of Americans ages twenty-five to fifty-four were in the labor force in September 2019, according to figures from the Department of Labor. That is one full percentage point lower than it was in the late 1990s.

Given this abundance of casualties from the creative destruction that is so championed by experts in business, US employers should have plenty of people to hire. Yet some of the United States' largest companies say the people on the sidelines of the economy are a mismatch with the skills set that companies now require.

Tim Cook, the CEO of Apple, revered in corporate and woke circles, falls into this camp. From his point of view, Apple is unlikely to find workers with the right skills among Americans now out of the workforce. Then again, Apple could hire workers away from other firms and train them. That would create openings in the jobs those workers left behind, and so on, eventually creating jobs for those out of the workforce.

Foreign workers will take less pay, not just because they might be from a country with less opportunity but because living in the United States is valuable in itself. Visas to work in the United States are a form of corporate welfare that enables businesses to pay workers less than they would have to pay a citizen to do the same job. The promise of a job that allows you to move to the United States is much more valuable to a foreign worker, no matter what the pay. And the size of the subsidy is enormous.

If foreign workers bear children while they are in the United States, the kids are considered to be US citizens, and eventually, they will be able to sponsor their parents for permanent US residence and citizenship.

Getting to live in the United States is part of the compensation foreign workers receive for accepting employment from a US company. Americans want to be compensated in dollars or benefits—forms of compensation that subtract from a company's bottom line. Immigration keeps labor costs down and improves the bottom line.

No wonder that, in the opening months of the Trump administration, the American Workforce Policy Advisory Board told him we have a worker shortage and urged him to abandon his policy of hiring Americans first. The panel includes fifteen CEOs and only one representative of the private workforce.

President Trump knows better. This is driven by their desire for cheaper imported talent overseas. Yet he, too, has been open at times to corporate chieftains' lobbying for looser rules and higher quotas on imported talent.

Two years into the new administration, a foundational view of the Trump Century started shifting. The argument that too much immigration of cheaper labor from overseas hurts US jobs and wages looked as if it might be softening. That would cause another uproar, this time on the right.

On December 20, 2018, on my show *Lou Dobbs Tonight*, I went on a bit of a rant that would rile the lib critics at the Mediaite website. Though I must say, they got it right, and it is what I have believed for thirty years. As reported by *Newsweek*:

Speaking on the Lou Dobbs Tonight *show on December 20 [2018],*
the host lamented how the U.S. has known about Chinese attempts
to steal secret information for decades, noting how there are allegedly
3,500 companies in China whose sole purpose is cyberespionage, but
"neither a Republican nor a Democratic presidential administration
has done a damn thing about it," reports Mediate. "Until now." . . .

"Hell, I can't understand why we wouldn't go to war over this kind
of monstrous theft," Dobbs said. . . .

"Frankly, I don't understand this," Dobbs responds. "Absent casu-
alties and that is killed and wounded, this is no different than Pearl
Harbor. I mean, we are watching the destruction of hundreds of thou-
sands, hundreds of millions and billions of dollars every year."

What I said.

In early 2019, President Trump was under siege by a daunting lineup
of controversies: Russiagate, the Mueller investigation, the feckless attorney general, Jeff Sessions, the firing of FBI director "Lyin'" James
Comey, China trade, tariffs, immigration, "kids in cages," the Fed
fight, Fake News, clashes with CNN. Are we exhausted yet?

All of these forces were swirling around the president at once. In
the BT (Before Trump) era, presidents made news a few times a week.
This president makes news a few times per day, almost every day, and
often on his favorite megaphone for shouting down the Fake News
media and end-running to get his message directly to his 80 million
followers: Twitter. Meanwhile, the president's advisors were working
through the details of immigration policy, and proponents of the same
old trade policies were much in attendance at the White House sessions.

On February 5, 2019, President Trump gave his third State of the
Union address to Congress. He made a departure from the speech's

script that was telling. The original words were supposed to be "Legal immigrants enrich our nation and strengthen our society in countless ways. I want people to come into our country, but they have to come in legally." Instead of saying that, he said this: "Legal immigrants enrich our nation and strengthen our society in countless ways. I want people to come into our country *in the largest numbers ever*, but they have to come in legally." (Italics added.)

John Binder, a journalist at Breitbart, the icon of conservative websites, would note three weeks later, "Spokespeople for the Chamber of Commerce, LULAC, George W. Bush Center, and Koch Industries dominate the immigration talks in the White House currently." LULAC is the League of United Latin American Citizens. Koch Industries, though it is demonized by the ultra-Left for climate change sins, opposes job immigration restrictions.

On March 2, President Trump appeared at the Conservative Political Action Conference (CPAC) at the Gaylord National Resort & Convention Center in National Harbor, Maryland, and spent two hours onstage talking about many things. The archconservative CPAC crowd is a favorite foil of the president. CPAC hosted one of the first political speeches he ever gave before running for president. Eighteen minutes in, he revealed a hallmark of the Trump presidency:

> *I don't know, maybe you know. You know I'm totally off-script, right? . . .*
>
> *And this is how I got elected, by being off script. [Applause.] And if we don't go off-script, our country is in big trouble, folks. Because we have to get it back.*

Twenty minutes later, he cited by name a *Washington Post* reporter who had shown up at a Trump rally four hours early, taken a photograph of the empty arena, and "put out a note [on Twitter]—something to the effect, 'Not very good crowd size, Mr. President.' And I never saw it because I don't follow the guy." In fact, thousands of people had attended the rally, and 25,000 fans of the president had thronged outside the venue.

"From the day we came down the escalator, I really don't believe we've had an empty seat," the president told the crowd, warming things up.

Forty-three minutes into the one-man show, the president prepared to get serious and back onto the teleprompter: "All right, now let's get back to what I'm here for."

Then, fifty minutes in, he dropped a trade bombshell. He did it in his imitable elliptical style, so that, if anyone in the crowd spotted the startling flip-flop and shouted an objection, it was drowned out by chants of "USA! USA! USA!" As if CPAC were an Olympic event. Here is what the president said:

> There will be some people in the room that don't like this. We're down to 3.7 percent unemployment—the lowest number in a long time. But think of this: I got all these companies moving in. They need workers. We have to bring people into our country to work these great plants. . . . This was not necessarily what I was saying during the campaign because I never knew we would be as successful as we've been. Companies are roaring back into our country, and now we want people to come in. We need workers to come in, but they've got to come in legally, and they've got to come in through merit, merit, merit!

The last line sent the crowd into a CPAC version of a frenzy.

Breitbart caught the sudden shift in Trump terminology and posted a story online with the headline "Trump Touts Legal Immigration System for 'Our Corporations' at Expense of American Workers."

Four days later, on March 6, Apple CEO Tim Cook made an

appearance at the White House for an economic event. As the president and the Apple leader sat together for a press conference at the White House, suddenly the president was on board: foreigners were urged to apply. As cameras clicked away, he said, "We're going to have a lot of people coming into the country. We want a lot of people coming in. And we need it." That was exactly what Cook and other CEOs of huge companies wanted to hear. The president added, "We want to have the companies grow, and the only way they're going to grow is if we give them the workers, and the only way we're going to have the workers is to do exactly what we're doing."

Breitbart went apoplectic. The headline was ugly, both typographically and otherwise:

TRUMP ABANDONS 'AMERICA FIRST'
REFORMS: 'WE NEED' MORE IMMIGRATION TO
GROW BUSINESS PROFITS

Ahead of the 2020 presidential election, President Trump is abandoning his prior "America First" legal immigration reforms to support increases of legal immigration levels in order to expand profits for businesses and corporations.

For the fourth time in about a month, Trump suggested increasing legal immigration levels. With Apple CEO Tim Cook sitting next to him at the White House on Wednesday, Trump said he not only wanted more legal immigration but that companies needed an expansion of new arrivals to grow their business. . . .

The comments are a direct rebuttal of the president's commitments in 2015, 2016, and 2017, where he vowed to reduce overall legal immigration levels to boost the wages of U.S. workers and reduce the displacement of America's working and middle class.

The Breitbart broadside triggered more than ten thousand comments on the story online, most of them expressing outrage. It reported that in 2017 the president had said that legal immigration levels must be cut back to "reduce poverty, increase wages, and save taxpayers billions and billions of dollars." He had backed a bill aimed at cutting one class of immigration by 50 percent, to half a million entrants per year. Now this.

Meanwhile, the mess at the Mexican border was getting worse. Caravans of thousands of immigrants streamed toward the border, swamping Customs agents. By late March, the situation was dire, something had to be said. On March 28, my show opened with my calling on President Trump to fire Kirstjen Nielsen, the secretary of the Department of Homeland Security, for her department's shabby handling of the ongoing border crisis.

As the PoliticalDog101 lib site put it later, "Dobbs later blasted Nielsen for her 'ignorance' and called the leadership of the US Border Patrol 'absolute morons.'" Quite accurate. I cited our border agents, awaiting orders on what to do yet no one was helping them:

> They are sitting there waiting for orders, waiting for somebody to hand them a solution? If this is what we have come to, if the quality of people in leadership in DHS from the secretary of the department on down I mean lets, you know, just literally put out welcome wagons. Pile them high because we're just going to consign tens of thousands perhaps millions of Americans to their deaths.

The DHS secretary would resign a month later. That evening, I had also urged the president to shut down the border until we quelled this crisis. At 11:23 the next morning, he threatened to do just that, telling his 77 million fans on Twitter:

The DEMOCRATS have given us the weakest immigration laws anywhere in the World. Mexico has the strongest, & they make more than $100 Billion a year on the U.S. Therefore, CONGRESS MUST CHANGE OUR WEAK IMMIGRATION LAWS NOW, & Mexico must stop illegals from entering the U.S. . . . .

With a continuation at 11:37 a.m.:

. . . . through their country and our Southern Border. Mexico has for many years made a fortune off of the U.S., far greater than Border Costs. If Mexico doesn't immediately stop ALL illegal immigration coming into the United States throug [sic] our Southern Border, I will be CLOSING. . . .

And another continuation at 11:43 a.m.:

. . . . the Border, or large sections of the Border, next week. This would be so easy for Mexico to do, but they just take our money and "talk." Besides, we lose so much money with them, especially when you add in drug trafficking etc.), that the Border closing would be a good thing!

This president responds when the right viewpoint comes to his attention. Four nights later, on my show, I heaped praise on the president for his early order restricting travel into the United States by aliens who had been in China within two weeks of their intended entry. It came on the evening of April 1, 2020, and it was entirely serious.

I said he had put the China order into place against the advice of many of his top public health experts who advise him every day. He had saved thousands of lives as a result and bought us time to marshal a response. Yet he had been met by condemnation from the Left because he wanted to keep a stay-at-home order in place. I declared that night that President Trump wanted to be certain that we were saving American lives. He has been proven right at every point. Yet there is no acknowledgment of that from the Left, nor from our public health experts.

That brought a quick rebuke from Media Matters for America, a lefty site that trolls conservatives for a living: "Lou Dobbs Attacks Health Experts for Not Praising Donald Trump's 'Leadership.'" Note the quotes around "Leadership," a snide snub.

Just five nights later, though, it was time to get the president's attention again, this time regarding his softening resolve on immigration and the damage it does to US workers. This has long been a hallmark of his most basic beliefs. On April 6, I posted a tweet to the president:

Protect American Workers: Lou urges @realDonaldTrump to stop importing foreign workers as millions of Americans are losing their jobs and going without a paycheck.

The president always manages a comeback somehow. A few weeks later he took the bold step of imposing a temporary ban on *all* immigration to the United States during the Wuhan pandemic, spawning new gales of outrage—although the reaction to it was less maniacally out of kilter than it had been at the start of his presidency, when his ninety-day travel ban sparked waves of anguish and agonizing; then again, the nationwide lockdown blocked much chance of organizing protestors to march against him.

On April 29, I tweeted this:

Protecting America: @PressSec says @POTUS is putting the American worker first after deciding to put a temporary pause on immigration into the U.S. during this pandemic.

Conflicting agendas were confronting President Trump. He wanted to secure the borders to stop illegal immigration and tighten legal immigration into the United States. He also wanted to protect workers and wage growth, yet he would fare better with the help of big business. He may have been trying to balance and intermingle those interests.

Buried in a Breitbart story was a possible clue as to why the president had just seemed to shift his position, saying he had "mentioned he wanted to end the process known as 'chain migration,' where newly naturalized citizens can bring an unlimited number of foreign relatives to the country, and the Diversity Visa Lottery, which admits 55,000 random foreign nationals from around the globe to the U.S. every year."

In the Reagan days, that would have been called "linkage." President Trump had just rallied support for an immigration crackdown, which the ultralibs would revile. Ending chain migration and the diversity lottery would tilt the flow of immigration toward better-educated and better-qualified people. Democrats pushed back, but not that hard, as long as Trump didn't stem the supply of future voters for the Dems.

It may be that President Trump used the optics of the Tim Cook photo opp for a twofer: woo big business with the let-'em-in sentiment, and distract media attention from his aim to end chain migration. Maybe he was asking corporate America to help him move forward. For conservatives upset by his turnabout on immigration, he could cite his effort to end chain migration.

The sides are so polarized in the immigration and jobs debate, and the views are so polemical, that facts get lost in the fray. It is a bizarre oddity of these ridiculous times that when debating immigration, we aren't even allowed to explore and discuss the real numbers. Do we really know for sure how many immigrants are here, both legally and illegally, how many are arriving each year, how many is too many in terms of the economy and in terms of diluting or obliterating American culture?

In the April 2019 issue of *The Atlantic*, beloved Never Trumper David Frum penned an article entitled, "How Much Immigration Is Too Much?" Woke Twitter exploded with people saying it was racist simply to ask that question. *The Atlantic* retitled the piece "If Liberals Won't Enforce Borders, Fascists Will."

Today in the liberal hotbeds of New York and New Jersey, an immigrant living there illegally can get a driver's license—but a worker at an election voting precinct dare not ask for it if the "undocumented resident" wants to vote, against the law. In the Trump Century, the immigration debate will be inescapable, and the terms of the debate have been set: fewer, more productive immigrants or ever more random immigrants?

Even the US Census, the once-every-decade attempt to make a record of every living person in the United States, has become divisive. It might be good to know solid, specific numbers for how many of us were born in the United States, how many immigrants loved our country enough to undergo the arduous process and long wait to become a US citizen, and how many of us are noncitizen immigrants. And how many "undocumented residents" are living among us.

In these fractious times, a Trump administration official can be hanged in the public square simply for asking questions like that. In March 2018, the administration unveiled plans to add a simple question to the US Census form: "Is this person a citizen of the United States?" It would spark another fight fraught with more charges of racism and xenophobia, this time more of the latter.

In fact, the effort would lead to another Trump controversy rapidly making its way all the way up to the Supreme Court, which would decide the matter in another bitterly divided ruling. That case is up next.

CHAPTER TEN

FORBIDDEN FACTS

O ne secret is more even closely guarded than the identity of the phony whistle-blower in the Ukraine scandal. It is the precise number of foreigners, both legal and illegal, living in the United States.

This set of forbidden facts is being withheld by the radical Democrats, RINOs, and the left-wing corporate media. The Trump administration moved in 2018 to shine a light on the true numbers by adding a simple question to the survey form for the upcoming 2020 census: "Is this person a citizen of the United States?" That single query sparked a fractious fight all the way up to the Supreme Court of the United States.

President Trump would end up losing a big round and still prevail.

Every ten years, the US government engages in the enormous undertaking of collecting and tabulating information on every single person in the country. The US Census is required by Article I, Section 2 of the Constitution. It is used to collect information about where

Americans live, their incomes and occupations, their racial and ethnic makeup.

But not since 1950 has the basic short-form census questionnaire asked the straightforward question: Are you an American citizen? It stayed on the longer questionnaire given out to only a small portion of households until 2010—when the Obama administration dropped it altogether.

When the Trump administration moved in March 2018 to restore the question to the 2020 census, opponents demonized the president with the usual: it was racist, xenophobic, fascist, pick any or all three of the above. The left-wing litigation machine, led by New York State, dragged the issue into the federal courts.

Liberal lawyers argued that the question would intimidate Latinos, even those who were legal residents and US citizens, which would result in undercounting the population. Opponents also charged that Republicans were hoping that would be the result, because they wanted to use the question to apportion congressional and state legislative seats by counting only citizens—as if that would somehow be a wrong thing to do.

Another reason they cited was that the president is a racist; his tweets had shown it, and the pattern of his public comments alone was enough to render his action on the 2020 census to be illegal and unconstitutional. He had, they said, no constitutional right to do what he was doing.

It is unfathomable that so many on the left so vehemently oppose the simple act of counting how many US-born citizens, naturalized citizens, and noncitizens live in the United States. Immigration will continue to be one of the most bitterly divisive issues of the Trump Century. If we have no clear idea of the makeup of the US population, how can we agree on how much immigration is too much?

This is the age of cell phone movement data, federalized driver's licenses that track every flight you take, social media companies that know what you want to buy before you do. But when it comes to how many noncitizens live in our country, it is the age of forbidden facts. The radical Left views Americans as too racist, xenophobic, and deplorable to know the truth.

This view of Americans is an outrageous lie. We are a generous and welcoming people, keenly aware that most of us descended from immigrants who settled in this land. Our openness to immigration is a long-running American tradition that many of us would like to see preserved.

It is easy to see why the facts about immigration have been banished from public discussion: they defenestrate the liberal, anti-American narrative that ours is a deeply racist nation that hates immigrants. The stats show that the United States is a welcoming country. If our nation were anything like what the fever swamps of the Left conjure up, we would not have let the number of foreign-born residents grow to more than 16 percent of the population, where it lies today. And the United States would be the last choice for the more than 150 million people worldwide who wish they could move here.

Clear majorities of Americans tell pollsters they think immigrants strengthen America, are hardworking, and have strong family values. In 2018, a poll conducted for CBS News showed that 70 percent of Americans agreed that "welcoming and accepting people from different cultures" is important to the American way of life.

At the same time, however, polls show that Americans want limits.

In a 2018 Gallup poll, 79 percent of Americans said we need to have secure borders, including 93 percent of Republicans, 80 percent of independents, and 68 percent of Democrats. Some 72 percent of the US population wanted immigration to decline or stay flat. Only a quarter of people felt we should have even higher levels of immigration.

When pollsters ask these questions, they rarely inform the

respondent how many immigrants we actually do let enter our country. This matters. Americans may be underestimating the scale of immigration to the United States.

In 2018, the United States granted 1 million foreigners lawful permanent residence. That was the start of the numbers. Almost 4 million more foreigners were granted temporary work permits. More than 22,000 refugees were admitted (down from nearly 85,000 in the final year of Obama), and 38,687 people were admitted under asylum provisions. Plus, almost 2 million foreigners were admitted on student visas.

That adds up to more than 7 million foreign newcomers admitted to live here, for some length of time, in a single year. Yet almost 70 percent of Americans say that immigration should stop at 1 million newcomers or fewer per year—only one-seventh of the number we actually let in annually.

In a Rasmussen poll released in December 2019, 36 percent of respondents said they want fewer than half a million immigrants per year, and 15 percent said fewer than 750,000. Also, 17 percent said they were happy at the 1 million mark, and 17 percent said they wanted 1.5 million or more.

In other words, the United States has been carrying on a years-long debate about immigration and our future, our values, and our population in the midst of incredible misinformation about the facts of immigration. It is almost as if the left-wing corporate media, the RINOs, and the Democrats want to mask the facts from their fellow Americans. They prefer to focus public attention on sentimental appeals to the famous poem at the base of the Statue of Liberty. (See chapter 11.)

If we decide our demographic destiny through democracy, we will have fewer immigrants. If the dictates of the open-borders crowd decide it, immigration will explode.

The new left-wing religion on immigration is profoundly irrational and implacable. Its adherents constantly deny that they are for open borders, but they fight against any attempt to set numerical limits on immigration or set new criteria. They oppose all practical attempts to enforce existing laws.

For any public figure to confront the realities of too much immigration and weak borders—lower wages and job prospects for Americans, a higher burden on entitlement programs, cultural dissension—is to court charges of racism. He or she is prohibited from articulating any limiting principles for immigration or any restrictions.

When you never know when to say "enough," you are committing yourself to the position of "more," now and always.

Here are some more forbidden facts about immigration. The year Donald Trump took office, 44.5 million immigrants resided in the United States, the highest number since census records have been kept. More than 9 million of them, or 21 percent, arrived in the United States after 2009. Another 11.6 million, or 26 percent, arrived after the turn of the century.

One in every seven people, or 14.3 percent, residing within US borders is foreign born, according to a Pew Research Center analysis of the Census Bureau's American Community Survey. That is the highest share of foreign-born people in the United States since 1910, when immigrants accounted for 14.7 percent of our population.

The newest arrivals are more likely to come from Asia than from Latin America, with East Asia and India now the largest sources of immigrants. The Pew Research Center says that 37.4 percent of immigrants in 2017 were from Asia, compared with 26.6 percent from Latin America. The Philippines, El Salvador, Venezuela, the Dominican Republic, and Cuba are major contributors. Mexico, which dominated immigration flows from the 1970s through the first decade of this century, has been sending far fewer across the border.

Add the 45 million US-born children of immigrants, and

immigration accounts for 28 percent of our population, according to the 2018 Current Population Survey. The Pew Research Center forecasts that the immigrant-origin share of the US population will rise to 36 percent by 2065.

We are uncertain how many illegal aliens live in the United States. For obvious reasons, illegal immigrants are reluctant to talk to census workers, which makes official figures unreliable. Pew says they account for about a quarter of all of the foreign-born population in the United States. That is roughly 11 million "undocumented immigrants," in the terminology now required by the word police, as if those people had somehow arrived without their hall pass.

The Federation for American Immigration Reform (FAIR) puts the total at 14.3 million people living here illegally, up more than 40 percent in ten years from 10 million in 2010. Include their US-born children, and their ranks likely add 30 million to our country's population. The top five countries of birth for illegal aliens are Mexico (53 percent), El Salvador (6 percent), Guatemala (5 percent), and China and Honduras (3 percent each), according to the Migration Policy Institute.

Those numbers would be far higher if not for the Trump administration's efforts to stem the tide of illegal immigration. From September 2018 to September 2019, US border authorities apprehended more than 1.1 million migrants who crossed the border illegally from Mexico into the United States, according to US Customs and Border Protection.

That was the highest level of border apprehensions since 2007, when border agents detained 858,638 migrants crossing our southern border. Increasingly, entire families are crossing onto US soil. In 2019, the border authorities detained 473,000 families at the border, nearly four times the number of the prior year. Apprehensions of unaccompanied children increased by 52 percent to 76,020.

Arrests at the southwest border peaked long ago at 1.62 million in 1986, the year Congress enacted the Immigration Reform and Control Act, a new law that granted legal residency status to almost 3 million illegal aliens living in the United States. After that, illegal crossings dropped by half, then started climbing back up in 1999 to reach an all-time high of 1.64 million in 2000.

The rise and fall track the US economy's rise and fall. The vast majority of illegal residents in the United States come here not to flee repression or violence but to seek better wages and a better life. Border arrests hit a forty-year low of 327,577 in 2011, when the United States still was hurting from the Great Recession of 2009. Then apprehensions jumped to 600,000 a year in mid-2019, rising by 201,497 in the first four months of the fiscal year.

The above figures are from the nonpartisan Congressional Research Service.

The total number of nonimmigrant admissions into the United States—that is, those who are officially permitted to stay only temporarily—was 186.2 million in 2018, an increase of more than 7 million from just two years prior. Around 65 million people came in as tourists. Nearly 9 million came in on business trips. Half a million were diplomats or representatives of foreign governments.

Most of the visitors go home as they promised to do, but when the numbers are this large, even a small amount of abuse can add up to big numbers. A Department of Homeland Security study of visa overstays in 2018 looked at a subset of the year's admissions that included 54.7 million temporary admissions of people who had been expected to depart that year through air or sea points of entry. It found an overstay rate of 1.22 percent, or 666,582.

This figure fails to include most of those who arrive legally from Canada or Mexico, because they enter via land border crossings. The Trump administration has cracked down on visa overstays, which likely means that this represents a decline from the Bush-Obama era when overstays were tolerated if not encouraged by lax enforcement.

Today it feels as if the Left wants to let in all of the many immigrants who would come to the United States if they could. "Give me your tired, your poor, your huddled masses yearning to breathe free." How many huddled masses, exactly? Perhaps we should do the kind and moral thing and let in all those who have a dream of pursuing freedom and a better life in the United States, the land built by immigrants. Say, an additional 200 million people or more? Added to the existing US population of 330 million? It is a plausible possibility, given the findings of a poll of citizens of the world by Gallup in 2019.

The United States is the number one destination of people around the world. Apparently, the rest of the world sees us as something better than the wretched land of racist, white privileged xenophobes depicted by the US native left wing. Gallup has estimated that 160 million adults in foreign countries want to come live in the United States.

Almost 200 million more people cite the United Kingdom, Germany, and a handful of other Western nations as their preferred destination. It is likely that many of them would choose the United States if our borders were utterly open and their first choice was unavailable. That means we could all but double our entire population in only a decade.

That counts only the adults. Add the children the new immigrants bring with them and the ones they would bear in the United States, and a hundred million more newcomers could swell the population— although at that point the United States of America would have ceased to be recognizable as the country it has always been. Good-bye, USA.

We wouldn't be *United*, and our federalist system of autonomous and independent states would collapse, so no longer would we be a nation of *States*. US-born Americans would form so small a part of the population that we might as well drop *America* from our country's name, as well.

Demographics is destiny—and demographics with open borders is doom.

Here is another forbidden question: How much crime is committed by foreign-born criminals in the United States? To ask this question in a public forum is deemed racist per se by the Left. Crime statistics count criminals by race, by gender, by age, and by whether or not they identify as Hispanic. But there is no reliable measure of the role immigration plays in crime in the United States.

The Trump administration in 2017 undertook a study to determine how many illegal aliens reside in federal custody. It found that a surprising 21 percent of the almost 40,000 inmates in the Federal Bureau of Prisons were illegal aliens. In the general US population, they make up only 4 percent of the total.

Ninety percent of jailed convicted criminals lie in state and local prisons and jails. If the same rate applied at those levels, it would mean that almost 200,000 illegals are incarcerated in the United States; as well as 42,000 are in immigration detention facilities, according to the Prison Policy Initiative.

That was, unbelievably, the first federal government study of its kind. It counted only nonresident criminals, leaving out legal permanent residents here on green cards and naturalized citizens who ended up in custody. The report found that the Department of Justice had 58,766 known or suspected aliens in its custody, with the Federal Bureau of Prisons holding 39,455 and the US Marshals Service holding 19,311.

To paraphrase Donald Trump on the day he and Melania glided down the escalator into the Trump Tower lobby to announce his run for the White House, it is clear that the rest of the world is not sending us their best.

The absence of reliable data on the noncitizen population in state and local prisons makes it difficult for the public and policy makers to accurately assess the kinds of crimes and how often they are committed by people who entered the United States illegally. One obstacle: sanctuary city policies. Many left-wing cities avoid collecting data on the citizenship status of those they arrest or convict, lest the data help federal authorities enforce our immigration laws.

On March 26, 2018, the Department of Commerce, the parent agency of the Census Bureau, announced plans to add the citizenship question to the survey. Instantly groups across the country filed lawsuits challenging the proposal. Startlingly quickly, the case climbed to the Supreme Court.

In federal district court in New York, opponents of the Trump plan for the census had challenged it on the grounds that it violated the Enumeration Clause in Article 1, Section 2 of the Constitution, which mandates that a nationwide census be conducted. Judge Jesse Furman, an Obama appointee, rejected that argument.

But he found another reason to side with the plaintiffs, one they hadn't suggested. Judge Furman, ruling from the bench in district court on Foley Square in lower Manhattan, decreed that adding the census question violated the Administrative Procedure Act. He decided that the reason cited by the Commerce Department that the question was needed to enforce the Voting Rights Act was bogus, a "pretext" to conceal the real motive.

On June 27, 2019, the Supreme Court issued its ruling in *Department of Commerce et al. v. New York et al.*, and it was just as convoluted as the lower court's decision. The high court held that it was entirely legal for the citizenship question to be asked on the census survey—but it blocked the Trump administration from asking it. That was as ridiculous as it sounds—and a condescending intrusion on the powers of the executive branch.

Even more absurdly, five Republican-appointed Supreme Court justices ruled that the Trump administration had the authority under the Constitution to add the question—as they should have. Then one of the justices switched sides and banned the question because he viewed the government's explanation for adding it to be insufficient.

The switch in nine that killed the question was Chief Justice John Roberts, the same Bush appointee who had saved Obamacare by siding with his fellow liberal judges. In his ruling, he chided the Trump administration for providing a "contrived" explanation in court for why it wanted the question added in the first place.

Once again, the chief justice had managed to clone himself to vote in two ways: he had ruled with the four other Republican appointees that asking the question was legal, and he had defected to the other side and the four Democratic appointees in ruling that it was wrong to ask the question. Pick a side, please, Mr. Chief Justice.

On the Obamacare case in 2012, he had cast a similar two-faced vote that had saved the program by reclassifying, as a tax, the financial penalty the Obama administration had sought to impose on any American who refused to enroll in the program.

The Roberts ruling in the census case brought a chastising dissent from Justice Clarence Thomas, who criticized the New York district judge for being "predisposed to distrust" the Trump administration and for creating "an eye-catching conspiracy web" from random facts. A law professor at the University of Baltimore, Garrett Epps, later wrote in *The Atlantic*, "As judicial conduct used to be measured, it was a shocking breach of protocol. Yet Thomas's opinion was joined by Trump's two appointees, Neil Gorsuch and Brett Kavanaugh."

On my Fox Business show the night of the court's ruling, I told my viewers that maybe it was time to consider civil disobedience. The Media Matters for America liberal watchdogs posted my words on their website, which was nice of them:

> *The Supreme Court's support for open borders is on full display. The so-called high court today blocked the Trump administration's citizenship question in the 2020 Census. Think of this. The Supreme Court ruled against the question, reached into the Census Bureau, and decided about what questions could be asked because of, quote, "incongruent reasoning." Much like Chief Justice Roberts did when he saved Obamacare by twisting a fine into a tax. Is it time for the Trump administration to outright defy the activist court and put the citizenship question in the 2020 Census? Who are these judges, these justices, to decide what questions can and cannot be asked by the Census Bureau? The courts continue to rule against the American people in favor of illegal immigrants.*

Later that day, I posted a follow-up tweet:

High Court rules against Citizens and Citizenship Question because of "incongruent reasoning" reasoning: like Roberts' opinion to save Obamacare by twisting a fine into a tax. By that standard, half of this Court's rulings would be pitched! . . . MAGA @AmericaFirst.

Five days later, on July 2, 2019, the Commerce Department put out a statement saying it had begun the print run on forms for the 2020 census. Without adding the burning question. New York representative Carolyn Maloney hailed the victory, putting out a statement saying the "ominous storm cloud over the census has been lifted."

Finally, the other side had won a fight with President Trump, and he had retreated. His foes were still unaware of the fact that President

Trump never retreats—though he does retweet. A day later, he sent the press reeling when he contradicted Commerce. He posted a tweet at 11:05 a.m. on July 3:

> The News Reports about the Department of Commerce dropping its quest to put the Citizenship Question on the Census is incorrect or, to state it differently, FAKE! We are absolutely moving forward, as we must, because of the importance of the answer to this question.

That so astonished CNBC that its website ran a three-line headline blaring:

> Trump says he is 'absolutely moving
> forward' with census citizenship question,
> contradicting his own administration

Cue standard backlash. This one was a threefer—the libs could brand it racist *and* xenophobic because of the citizenship question, and they could stake new ground on the dictator/fascist front at the same time. In their view, President Trump had blatantly threatened to defy the law of the land as laid down by SCOTUS.

Perhaps they had misread the president's message, although, admittedly, the man can be imprecise at times. He had said that the "quest" for adding the question continued. That might turn out to be a different thing entirely. A week later, he backed down, or so it appeared. The *Washington Post* ran a story online at 8:30 p.m. on July 11 with the headline "Trump Retreats on Adding Citizenship Question to 2020 Census." The article began:

> *President Trump on Thursday backed down from his controversial push*
> *to add a citizenship question to the 2020 Census, effectively conceding*

defeat in a battle he had revived just last week and promised to continue
despite a recent string of legal defeats.

The *Post* article noted that after Commerce had made plans to ex-
clude the citizenship question from census forms, "a furious Trump
reversed that decision the next day, saying that he was not giving up
on asking about citizenship." He had said that the "quest" to add the
citizenship question would continue.

And it did. The president's people simply looked for another way to
gather the same information. They found it a day later. While Trump
haters and the Democrats celebrated their seeming victory, President
Trump signed Executive Order 13880. It requires the US Census Bureau to produce data on the citizen voting-age population by the end
of March 2021. It also orders all other relevant agencies to share their
databases with Census. This shall be done "with the goal of obtaining administrative records that can help establish citizenship status for
100 percent of the population." At the very top of the new order, in Section 1, the president began by citing the Supreme Court ruling that had
just gone against him. The ruling did, after all, say that it was entirely
legal to add the question the Democrats had so reviled and feared.

The executive order, usually a formal, stilted document, was in this
instance startlingly transparent about its intentions. It was combative
about the Supreme Court ruling and almost conversational in relaying
the president's frustration. It noted that the high court had ruled that
the Department of Commerce can "lawfully include" the question
on the census form "and, more specifically, declined to hold" that the
decision to do so was "substantively invalid."

The tone is almost acidic:

That ruling was not surprising, given that every decennial census from
1820 to 2000 (with the single exception of 1840) asked at least some

respondents about their citizenship status or place of birth. In addition,
the Census Bureau has inquired since 2005 about citizenship on the
American Community Survey—a separate questionnaire sent annually
to about 2.5 percent of households.

The order added, "I disagree with the Court's ruling" and said
that it "has now made it impossible, as a practical matter, to include a
citizenship question on the 2020 dicennial census questionnaire." So
President Trump, once again, had found a way around the obstacles to
move forward in doing what he wanted to do. His "quest" continues.
He had found a way to reach his goal without risking another media-
fed, fake constitutional crisis.

Overall, the Trump administration was racking up an impressive
record of wins in court. The fight over the 2020 census showed that
President Trump's remaking of the federal judiciary already was having beneficial effects. Incumbent liberal judges were throwing obstacles into the administration's path, and they were being overturned,
sometimes directly by the Supreme Court.

The president's two appointees, Neil Gorsuch and Brett Kavanaugh,
were voting with the administration, as he had hoped. The forty-four
appeals court judges already named during President Trump's reign
now occupied fully one-quarter of the circuit bench, providing him
with more backup. And with more than a hundred newly named
judges at the federal trial level, his odds of winning the opening round
were going up.

In July 2019, the Supreme Court had endorsed his authority to use
$2.5 billion in Pentagon funding to start building the border wall and
cleared a new policy barring most Central American migrants from
seeking asylum in the United States. It also ruled in favor of the administration's right to ban transgender people from joining the military, thwarted the effort by Trump's enemies to reveal his tax records,

and okayed the Trump plan to deny a green card to any immigrant who might end up on the government dole.

The census case had leapt up to Supreme Court review with surprising speed, thanks to a new tack the Trump administration had been taking, to the delight of conservatives and the consternation and fretting of legal scholars on the left.

President Trump may have set a record for how many cases he has pursued in the high court in only three years as president, so much so that a sitting justice would call out the Republican majority on this in a dissenting opinion in one of the rulings—and draw fire from President Trump on Twitter.

Usually, when a federal district court judge rules against an administration, lawyers must file an appeal to the next level up, the circuit court of appeals. A dozen appeals courts oversee rulings from the lower district judges. But Trump lawyers often skipped the circuit appeals level entirely, filing an emergency appeal straight to the US Supreme Court to ask it to lift injunctions filed by district judges two levels down.

The administration used the tactic thirty-two times in the first three years of the Trump presidency—compared with only eleven times in eight years of the Obama administration. In other words, Trump pulled the shortcut eleven times a year, compared with only 1.3 times a year for Obama.

The Trump team does this in three ways: it asks for an emergency stay; it petitions the high court to bypass lower courts and move a case straight onto its own docket; or it takes the rare step of petitioning the Supremes for a writ of mandamus, a mandatory order instructing a recalcitrant lower court to follow orders it has refused.

Overall, the Trump lawyers requested twenty-one emergency stays of lower-court orders (eight in eight years for Obama), nine petitions for review (three for Obama), and three writs of mandamus (zero for Obama). Many of them succeeded.

In the first months of 2020, President Trump racked up two more immigration policy wins at the Supreme Court. In January, the justices ruled 5–4 to overturn the nationwide injunction ordered by a federal district court judge in New York. That allowed Trump to proceed with plans to deny green cards to immigrants who stood a chance of ending up on food stamps or other entitlements. The case had bypassed the Ninth Circuit Court of Appeals in San Francisco.

Garrett Epps at the University of Baltimore, who published those numbers in a legal abstract in September 2019, noted the Supreme Court's "troubling pattern of deference to Trump's wishes. . . . The government's use of these procedures smacks of entitlement, of a sense that Republicans went to great trouble to tilt the Court in their favor and should now reap their reward."

In February, the justices ruled in a related matter, voting 5–4 again to vacate an injunction, this one imposed by a district court judge in Illinois. The lower-court injunction covered only the judge's district. That ruling let the Trump administration skip past the Seventh Circuit Court of Appeals in Chicago, prompting a stinging seven-page dissent by Obama appointee Justice Sotomayor, who chided the Court's Republican majority:

> *Today's decision follows a now-familiar pattern. The Government seeks emergency relief from this Court, asking it to grant a stay where two lower courts have not. The Government insists—even though review in a court of appeals is imminent—that it will suffer irreparable harm if this Court does not grant a stay. And the Court yields. . . .*
>
> *Claiming one emergency after another, the Government has recently sought stays in an unprecedented number of cases. . . . And*

with each successive application, of course, its cries of urgency ring increasingly hollow.

Justice Sotomayor, joined in her dissent by Justice Ginsburg, added, "Perhaps most troublingly, the Court's recent behavior on stay applications has benefited one litigant over all others." That so angered President Trump that he threw down about it on Twitter even as he was on a trip out of the country. While in New Delhi, Trump tweeted to my Fox colleague Laura Ingraham:

> Sotomayor accuses GOP appointed Justices of being biased in favor of Trump. . . . This is a terrible thing to say. Trying to "shame" some into voting her way? She never criticized Justice Ginsburg when she called me a "faker." Both should recuse themselves. Both should recuse themselves on all Trump, or Trump related, matters!

President Trump was referring to public remarks made by Justice Ruth Bader Ginsburg about Trump while he was running for office. She later apologized and acknowledged that a justice never should say something like that. No doubt her enmity remains. If he ever had managed to browbeat them both into recusal, President Trump would suddenly have had a 5–2 supermajority on the high court, and even have-it-both-ways Chief Justice Roberts would be unable to screw that one up.

Even after the Supreme Court ruling in the census case, the controversy lingered. By March 2020, the administration had canceled census fieldwork, after 70 million people in the country had been surveyed. Now, on April 13, Secretary of Commerce Wilbur Ross and the director of the Census Bureau, Steven Dillingham, filed a formal request to Congress, asking for a delay of 120 days. That would delay the reporting of data that would determine the realignment of congressional seats.

The need was clear; the US Census relied on online surveys and in-person interviews and home visits, and the in-person work had been halted for a couple of months. Instantly, however, Trump foes presumed that something darker must be at work. Maybe he was secretly planning to use a delay to manipulate the results to favor Republicans in congressional reapportionment. Or he might have been looking for a backdoor way to frighten illegals and even legal foreign residents.

When the request for a delay in the census survey went to the House Committee on Oversight and Reform, its Democratic chairwoman, Representative Maloney of New York, again reacted with suspicion. She put out a chilly statement saying that the panel would consider the request, but "the Trump administration was stonewalling in providing information vital in assessing the move," as a *Politico* story put it a few hours later.

The Democrat chided the Trump team for putting Secretary Ross on a call to congressmen without including the Census Bureau director. Her committee had been unable to get a briefing from him. She put out an indignant statement: "If the Administration is trying to avoid the perception of politicizing the Census, preventing the Census Director from briefing the Committee and then excluding him from a call organized by the White House are not encouraging moves." But the US Census, by that time, had been politicized long before, and the Left had done it.

That petty exchange played out as the Wuhan crisis was beginning to gather momentum. In the coming months, as the pandemic turned into the biggest challenge our nation has faced since the Great Depression, the Trump resistance pressed on. It pushed even harder when we should have dropped the bitter infighting and come together to slay a common invisible enemy.

The fight over a simple question on the 2020 census was a kerfuffle unworthy of all the ink and pixels it received. As President Trump had put it, speaking at an afternoon event in the Rose Garden on the day the government had begun printing US Census forms without the burning question, "I'm proud to be a citizen. You're proud to be a citizen. The only people that are not proud to be citizens are the ones who are fighting us all the way about the word, 'citizen.'"

CHAPTER ELEVEN

DEMONIZING A PRESIDENT

The touchstone of all Trump policy was a clear, immutable concept: America first. It drove everything President Trump did, and for a US leader the principle should be unassailable. It was difficult for him to get that message across to Americans above the constant drone of angry accusations from embittered foes, who argued his real agenda was driven by hidden intentions that were racist and xenophobic.

This constant suspicion cast a pall over even a sensible-sounding proposal when the issue involved immigration and how to limit or refine it. In August 2017, Trump senior advisor Stephen Miller held a press conference to discuss plans to reform immigration in ways that could make newcomers to the country a better fit.

By that time Miller was already a target of critics. He had played a pivotal role in crafting strategy for a hat trick of Trump outrages: the travel ban, a reduction in refugees, and the policy approach that Never Trumpers would come to call "kids in cages," referring to the

separation of children from their parents when families were apprehended crossing the border illegally.

On the initiative to cut back on refugee admissions, the Trump administration had made progress, though the libs would view it as more
racism. In fiscal year 2018, total refugee admissions were reduced to
45,000, down more than half from the last year of the Obama administration, when 110,000 refugees had been allowed to settle in the
United States. The plan was to reduce the number further to 30,000 in
2019 and to 18,000 in fiscal 2020. Prior to 2018, the US had resettled
more refugees each year than all other nations combined. Combined!

Now Miller was briefing the media on new proposals that would
give preference to immigrants who speak English and are highly educated or who possess high-tech skills that are in low supply in the
United States.

The loose guidelines stopped short of blocking anyone from im-
migrating to the United States. They offered points in favor of some
immigrants over others when deciding which of any two applicants
should be admitted. That managed to offend CNN's White House
correspondent, Jim Acosta. Acosta was supposed to be a fair and objective reporter but long before had betrayed himself as a dim Dem
partisan. The reality was that he hadn't asked a question in years that
made news about anything except Jim Acosta.

Acosta, sitting among the reporters at Miller's briefing, began,
"What you are proposing, what the president is proposing here does
not sound like it's in keeping with the American tradition when it
comes to immigration." Then he revved up a rhetorical run: "The
Statue of Liberty says, 'Give me your tired, your poor, your huddled
masses yearning to breathe free.' It doesn't say anything about speaking English or being able to be a computer programmer."

Miller was ready for him. He countered that our immigration system already has an English-speaking requirement for those who apply

for citizenship. He is right: they must pass a proficiency test of their ability to read, write, speak, and understand English, as well as a knowledge of US history and government, based on a civics test that many Americans would be unable to pass. Giving applicants for US residency an advantage for speaking English would only refine the system that was already in place before the Trump takeover.

Miller also told the reporters watching the faceoff that the Statue of Liberty, a gift from the French to the United States to celebrate our centennial in 1876, had arrived without the famous inscription. The inspiring line had been taken from the Emma Lazarus poem "The New Colossus," composed in 1883 as part of the efforts to raise funds for the pedestal for the statute's base on Liberty Island.

Miller told the crowd: "I don't want to get off into a whole thing about history here"—a clear sign that he was about to do exactly that—"but the Statue of Liberty is a symbol of liberty and lighting the world. It's a symbol of American liberty lighting the world." That's a Trump technique, repeating a key phrase twice. Okay? Twice. Yes, I meant to do that.

He went on, "The poem that you're referring to, that was added later, is not actually a part of the original Statue of Liberty."

Acosta replied, "That sounds like some National Park revisionism. . . . The Statue of Liberty has always been a beacon of hope to the world for people to send their people to this country." He sounded a little angry or embarrassed. Or maybe he was grandstanding.

Miller's cogent reply highlighted the ungrounded fanaticism of typical Left claims that a change in our laws or a reduction in immigration was a betrayal of America and the giant copper lady in New York Harbor. (Her pale green exterior is a patina from 150 years of standing as a "beacon of hope.") Miller, in a rant of his own, called out Acosta by name:

In 1970, when we let in 300,000 people a year, was that violating or not violating the Statue of Liberty? . . . Tell me what years meet Jim Acosta's definition of the Statue of Liberty home law of the land? So you're saying a million a year is the Statue of Liberty number? 900,000 violates it? 800,000 violates it?

Acosta retreated into the last resort of the Left, or these days maybe it is the first resort: accusations of racism. "Are we just going to bring in people from Great Britain and Australia?" he asked. After a few more minutes of on-camera debate, Acosta again: "It sounds like you're trying to engineer the racial and ethnic flow of people into this country."

Acosta has abundant company in combining a lack of knowledge of immigration policy with maniacal adherence to an ideology with no basis in US history. He had just had the misfortune— or the pleasure, in his eyes—of reciting the catechism of open borders on live television.

Maybe Acosta's remarks could have been called racist, too. The idea that only immigrants from Great Britain and Australia speak English is laughable. In recent years, the United States has accepted arrivals from more than twenty other countries where English is the official or dominant language.

This list includes Caribbean nations such as Barbados, the Bahamas, and Jamaica. We have accepted immigrants from English-speaking African countries such as Liberia and Zimbabwe. People all over the world speak English. In fact, two-thirds of immigrants to America speak English well, even under our system that gives no explicit preference for that skill. Less than 14 percent speak no English at all.

The demand that US policy be based upon a genuflection to the Statue of Liberty and the line from "The New Colossus" is a sure sign of immigration fanaticism. It is worth recalling that the basic symbols of the US political tradition—the Declaration of Independence, the

Constitution, and the Bill of Rights—were adopted through democratic processes.

The representatives of the colonies at the Continental Congress debated, deliberated, and eventually approved the Declaration of Independence. The Constitution was approved at the famous convention in Philadelphia and had to be okayed by the states before it could become the law of the land. The first ten amendments to the Constitution that have come to be called the Bill of Rights were adopted by the House and Senate and then ratified by the states.

These are the true laws of the land—not "The New Colossus," a socialist's poem.

This reverence for the leftist reading of the Lazarus poem is a very recent invention. When it was attached to the Statue of Liberty, few would have argued that the United States had a moral obligation to take all comers. Few would have said that the immigration policy then in existence was inviolable. For many immigrants over the decades after the Statue of Liberty arrived in New York Harbor, Lady Liberty stood as a beacon of hope and a sign that they had succeeded in their efforts to arrive in the United States and win entry to this country. Legally.

An open-door policy was never the idea. Indeed, nearly all Americans then and now would say that the primary driver of immigration policy should be the needs of the American people. America first. Those two powerful words, a hallmark of Donald Trump's strategy, invigorate and illuminate everything that is right.

In October 2019, Simon & Schuster, a longtime subsidiary of CBS, published a scathing book on the immigration controversies of President Trump, written by two reporters for the *New York Times*: *Border Wars: Inside Trump's Assault on Immigration*. In passage after passage, the authors, Julie Hirschfeld Davis and Michael D. Shear, exemplified the new fanaticism.

"Was it racism? Nativism? Xenophobia?" they asked, referring to the Trump administration's border policies. "Trump and those who knew him best swore that it was not. But Trump's instincts clearly tended toward bigotry—the belief that foreigners were a threat, and that native-born Americans were inherently more deserving."

Imagine that. Any belief that some foreigners could pose a threat was evidence of bigotry in the minds of the *Times* reporters, at least when they were examining Trump. They viewed it as bigoted that the US president would regard the American people as inherently more deserving of the benefits of living in the United States.

When advocates of even more immigration to the United States depict as racist almost any effort at reform or stepped-up enforcement, it hinders the effort to find rational solutions that can work. This causes gridlock, neutralizing any crackdown, and then any effort to ease up on restrictions is an open invitation to immigrants to come on down.

The Obama administration discovered this effect in the last year of its eight-year run, when it attempted to initiate a policy for letting first-time illegal entrants slide without penalty and prosecuting only repeat offenders for felony charges. Others would be detained and possibly deported, but never criminally prosecuted. That would encourage a surge in first-time border crossers.

The result was a huge influx of asylum seekers at the southern border, and it caught the Obama administration off guard. Undocumented and inadmissible migrants started showing up in bigger numbers. Many believed—correctly—that they would receive more lenient treatment if they presented themselves as "families." So migrants were increasingly showing up with children in tow. Sometimes the kids weren't theirs.

The Obama administration responded by setting up new detention centers to hold immigration detainees in fenced-in areas for security purposes. Officials had hoped that the detention centers would deter future undocumented aliens from showing up at the border. Instead, they may have encouraged them. Having your family gathered together under US guard behind a security fence has to be preferable to being back at home in Central America dodging death squads.

Then waves of unaccompanied children began arriving at the border. Many in Latin America had come to believe that children would be released rather than being sent back home alone. The Obama administration began holding "unaccompanied minors," an important term, in facilities separate from those of adults. This applied especially to children who had arrived with adults believed to be criminals, including child-sex traffickers.

In other words, the Obama administration was separating children from their families. It had adopted this policy first, not President Trump. Nor could the Obama people be faulted for doing so; they had been following the law. Detaining and processing kids separately from adults—even their parents—was required by a federal court order.

What about the fenced-in areas? Two years later, they would create the infamous "kids in cages" uproar and set off another mudslide of liberal outrage and race baiting. Left-wing immigration groups had complained about them in the Obama years, but the left-wing corporate media had mostly ignored the complaints. Perhaps they wanted to avoid making Obama look bad or avoid being called racist.

In 2016, a federal judge declared that neither accompanied nor unaccompanied minors could be detained. In fact, the district court held that their parents could not be detained, either. That last part was overturned on appeal by the Ninth Circuit Court of Appeals in San Francisco, which ruled that the government could hold the parents but had to release the children. The Obama administration decided it was unwilling to separate families in that way and, for the most part, stopped family detentions.

Instead of detention, Obama turned to a pilot program intended, as an NBC News story put it later, to "keep asylum seeking kin together, out of detention, and complying with immigration laws. It was praised by immigration advocates for both its high rate of compliance and its ability to help migrants thrive in a new country."

The administration's new approach was part of its Family Case Management Program. It should have been called "catch and release," as President Trump and others would later nickname it. Apprehended parents and their children would be told to show up for a hearing that wouldn't occur until months or even years later. Often they bailed on showing up at all. Finding out how often this happens can be like trying to unearth some state secret.

It was a recipe for unlimited open immigration. The law said that illegal border crossers and inadmissible asylum seekers should be returned home or prosecuted. The new Obama program did neither. A new policy aimed at helping illegals "thrive in a new country" was, in fact, facilitating more illegal immigration.

It was contrary to the spirit of our laws and to the interest of the American people. We have the right to control our borders and decide our immigration policy based on democratic deliberation, rather than just allowing those who show up on our doorstep, uninvited and unauthorized, to stay in this country indefinitely without penalty.

Because the federal courts had ruled that it was illegal for the government to hold children in the same place as adults, the kids had to be relocated in twenty days. Where to put them? The children were classified as "unaccompanied alien minors," as they had been in the Obama era, and were transferred to the Office of Refugee Resettlement (ORR) for care and custody, as the nonpartisan Congressional Research Service later reported.

Thus, even before President Trump took office, US government policy was to "separate children from their parents" and move them to another site; they were handed off to the custody and care of ORR, a part of the Department of Health and Human Services (HHS).

That was in keeping with the existing federal court order. Eventually, the children were to be reunited with Mom and Dad, and the family would be either returned home or allowed to establish legal residence in the United States. It also happens to be the policy of local police in the United States when they arrest, say, a father for drunk driving and his young kids are with him. The kids are placed in child services if a relative is unavailable to pick them up.

The Obama administration received little pressure from the adoring media for its policy of separating children from their parents when processing illegal immigrants. That changed hugely when the Trump administration inherited the policy and began fiddling with it to end "catch and release."

President Trump had warned his closest advisors that the Obama "catch and release" program would lead to disaster. But apprehensions at the border fell by 25 percent in his first full year in office, to 300,000 from 400,000 in the last year of the Obama administration. Obama holdovers and career bureaucrats inside the Department of Homeland Security resisted the Trump crackdown. For more than a year, patriots in the Trump administration sought to put a new system into place but got nowhere. As the number of illegal crossers with children rose to the highest level of Trump's presidency, it became clear that something more must be done.

Kirstjen Nielsen, the director of Homeland Security, resisted imposing the new policy for months. She had been tapped for her post in part because of her close relationship with the president's chief of staff, General John F. Kelly. Finally, a full year after the zero-tolerance regime was announced, Director Nielsen signed on in May 2018.

On May 7, the US Department of Justice unveiled a new

"zero-tolerance" policy on illegal border crossings, vowing to prosecute all adult aliens apprehended while crossing the border illegally—with no exceptions for even asylum seekers or those with young children. Asylum seekers could continue seeking legal entry into the United States.

Instead of catch and release, letting first-time illegal entrants pass without penalty as in the Obama era, the Trump administration would detain and prosecute all border crossers, first-time offenders and repeaters alike. The aim was to present a tough message to would-be immigrants: you will be arrested and held if you try to enter the United States illegally.

Indelicately, Trump officials said it directly. Attorney General Jeff Sessions declared, "If people don't want to be separated from their children, they should not bring them with them. We've got to get this message out. You're not given immunity."

Chief of Staff Kelly told NPR that family separation "could be a tough deterrent," instead of pointing out that federal law and a court order *required* the Trump administration to separate kids from their parents. It sounded punitive when it was, instead, necessary.

That fed the media frenzy already under way. The Trump administration had sought merely to enforce an existing law that had been cast aside in the Obama era, rather than seeking to revise an existing law or regulation. The nonpartisan Congressional Research Service pointed that out in a report in February 2019, adding, without judgment, that "Prior Administrations prosecuted illegal border crossings relatively infrequently."

Thus, the family separations were the unintended *consequence* of the zero-tolerance policy rather than its explicit objective. The media never reported that. Covering the outraged statements and all the fury and flurry of the Left was story enough for them. "Kids in cages," with its catchy alliteration and conjoining of innocence and the perverse, was too good to resist.

At one point, Twitter and the rest of social media lit up with a photo of a little Hispanic boy, his tiny hands gripping two thin bars of a cage, staring out off camera and sobbing. It was heartbreaking. It evoked millions of words in response online.

The photo went viral with a kicker: "Are you Trump fans really OK with this?" One of the earliest places it showed up was in the Twitter feed of a journalist/activist, Jose Antonio Vargas, who is also an illegal resident of the United States. He posted the photo on Twitter in a tweet that went out on June 11, 2018:

> This is what happens when a government believes people are "illegal."
> Kids in cages.

Thousands of people shared it with their followers, including the liberal actor Ron Perlman and other Hollywood figures.

It was a fake. Maybe not intentionally, but it was. Later it turned out that the photo had been taken at a rally in Dallas, far from the US-Mexico border. Protestors had set up a mock cage at the site. The little boy in the photo was the son of someone at the event. The child had walked inside the mock cage and turned around, spotted his mommy far away, and started crying. Picture snapped.

That was later documented by FactCheck.org. The original poster said he had been unaware of the photo's origins and had posted it to Twitter anyway—after which, of course, President Trump capitalized on the hoax on Twitter.

The media devoted so much effort to the "kids in cages" story that it felt as though tens of thousands of children were caught up in the system. In cages. A total of 4,337 immigrant children were held under the Trump zero-tolerance program from April 19 to May 5, 2018. They were spread among a hundred DHS sites, the CRS report stated. Several thousand more had been separated prior to the public announcement of the policy change.

Any child's being removed from a parent is a sad thing. Still, this was an avalanche of hysterical coverage for 4,000 children, out of more than 1 million people crossing into the United States illegally each year.

After the fracas, President Trump issued an executive order on June 20, 2018, requiring DHS to hold aliens awaiting criminal trial or immigration proceedings. As a result, DHS and Customs and Border Protection (CBP) stopped referring most illegal crossers for criminal prosecution. A federal court judge then ordered that all children in the system must be reunited with their families. A second judge rejected DOJ's request to modify existing rules to extend the twenty-day child detention guideline.

The same Democrats who condemned that flawed government program also argue that government is ready and able to take over the entire $3.5-trillion-a-year US health care system without a snag.

The Fake News media worked hand in hand with the Democrats on "kids in cages." They are one and the same. Liberals talk about the right-wing media and point to Fox News, Breitbart, Rush Limbaugh, and Mark Levin, but they run out of much to say thereafter; lib media are everywhere. They all seem to speak with the same voice, as if it were orchestrated. In the past, in fact, it has been orchestrated.

A decade ago, reports exposed the existence of a secret, private Google chat circle called JournoList, where four hundred liberal journalists and commentators chatted online, out of view of any public scrutiny. Some of the members of that chat group remain leading members of the media today, ostensibly as objective journalists covering the Trump administration.

The posts from mid-2008, as Obama was running for a first term, reveal how much liberal journalists hate this country, and how much they despise conservatives as being beneath them. The list was organized by Ezra Klein, a writer for the *Washington Post*, who later cofounded the lib website Vox, where he is still editor at large. It included this post from a writer for a nonprofit website, Spencer Ackerman, who today is the national security correspondent for the Daily Beast: "What is necessary is to raise the cost on the right of going after the left. In other words, find a rightwinger's [*sic*] and smash it through a plate-glass window . . . to let the right know that it needs to live in a state of constant fear." He suggested confronting conservatives "Fred Barnes, Karl Rove, who cares—and call them racists. Ask: why do they have such a deep-seated problem with a black politician who unites the country? What lurks behind those problems? This makes *them* sputter with rage, which in turn leads to overreaction and self-destruction."

A writer for *The Nation*, Katha Pollitt, spoke of how hard it had been to defend President Clinton during the Monica Lewinsky flap: "Let me tell you it was no fun, as a feminist and a woman, waving aside as politically irrelevant and part of the vast rightwing conspiracy Paula, Monica, Kathleen, Juanita." Though she managed.

Ackerman empathized with her: "But what I like less is being governed by racists and warmongers and criminals."

Also on JournoList was a *Nation* editor who is now an anchor on MSNBC, Chris Hayes. He urged colleagues to ignore the controversy over candidate Obama's longtime ties to a black reverend with a record of giving anti-American and anti-Semitic sermons. After all, he pointed out, "Our country disappears people. It tortures people. It has the blood of as many as one million Iraqi civilians—men, women, children, the infirmed [*sic*]—on its hands."

It has gotten worse now that this mob of radical-left journalists is covering President Trump. For any liberal critic of the president, slinging the racist tag is sufficient to make news, get quoted in the left-wing press, and make yourself a hero in the woke social media. They should send him a thank-you note.

Does it seem as though people in Hollywood have more opinions on politics than ever? Robert De Niro. Rob Reiner. Kathy Griffin (the redheaded "comedian" who posed for a photo shoot gripping a fake severed head resembling that of Trump). Chrissy Teigen. Alyssa Milano. Or is it that journalists have discovered that they can cover their opinions as news, sneaking into their articles the propaganda they could never express directly themselves?

When you choose to view everything through the lens of racism, you can find it anywhere you like. Wanting to tighten immigration standards to protect jobs is viewed as racist and antiforeigner. The same goes for reducing H-1B visas that let foreigners occupy US jobs on the cheap. Ordering a drone hit on a terrorist is equivalent to attacking a person of color. Pressing our allies in Europe to pay their fair share of NATO defense, after decades of welching, means that President Trump is a xenophobe who is naive about world affairs.

President Trump's standing up to China on trade was aimed at stoking US growth and protecting jobs. The left depicted it as anti-Asian. The same goes for closing down entry to this country by people who had visited China recently, in the first weeks after the Wuhan virus started spreading—and weeks after that, his shutting down all immigration temporarily, while the government was cauterizing the virus crisis. Xenophobic!

Even the Wuhan virus was racist: people of color died more frequently. It doesn't matter that minorities (has that term been banned?)

are more likely to live in densely populated urban areas in multifamily housing and more likely to take crowded mass transit. That has nothing to do with color. Still, the president must be at fault somehow.

The Dems have used the racism cudgel for twenty years to fight reform efforts for immigration and even free trade.

In the meantime, the opposition fails to offer a plausible answer to a couple of questions I presented on the topic years ago. How can you have freedom and lawful immigration if you fail to control immigration? How can you control immigration if you fail to control our borders? Never has anyone on the left offered good answers to these questions.

My stance has been steadfast on trade and immigration for more than twenty years. President Trump and I walked separate paths toward the same beliefs. It starts with America first. That applies especially to free trade.

All along, disciples and shills for the free-trade movement have vilified me for these views, with one-tenth the intensity they would fire at President Trump. More than a decade ago, when I hosted a nightly news and opinion program on CNN, I published an op-ed on the network's website, revealing some of this static.

My calls for a balanced US trade policy—to require our trading counterparts to buy as much from us as we buy from them—prompted a *Financial Times* editor to call me the "high priest of demotic sensationalism." "Demotic" means "of the common man," so perhaps that is what he meant to write; or was it "demonic"? An editorial in *The Economist* accused me of embarking on "a rabidly anti-trade editorial agenda" and "greeting every announcement of lost jobs as akin to a terrorist assault." A *Washington Post* columnist and American Enterprise Institute fellow, James Glassman, accused me of being "a table-thumping protectionist."

An opinions editor at the *Wall Street Journal*, Daniel Henninger, excoriated me in high style: "It's as if whatever made Linda Blair's head spin around in 'The Exorcist' had invaded the body of Lou Dobbs and left him with the brain of Dennis Kucinich." Now, that was clever and quite funny. It was a quadruple toe quip: an attack on me (one) and my argument (two) with a great movie reference (three) and a totally gratuitous sideswipe against a Democrat (four). Dan, a fine gentleman, is still in place at the *WSJ*, still espousing the views of the free traders who still oppose the Trump agenda.

My column at CNN.com back then added a point that resonates more than ever as we embark on the next decade:

Those quotes are from some of the most respected news organizations, and there have been dozens of other articles critical of my view that outsourcing American jobs is neither sound, smart, humane, nor in the national interest.

I will tell you it does make a fellow think when attacked so energetically and so personally. But in none of the attacks on my position on outsourcing has a single columnist or news organization seen fit to deal with the facts.

This must be how President Trump feels today and has for much of his presidency. He had to wrangle with a far wider circle of enemies and on so many more policy fronts where he was trying to change the way things had operated forever.

As divisive as the debate has been in America, with sometimes bizarre moments in which victims real and fake are elevated into talking points, one wishes all could agree on President Trump's central ethos: America first. It is a simple rule of survival. No nation ever should put the interests of another nation ahead of its own.

It harks back almost two hundred years to the concept of American exceptionalism, which drove the country's sense of destiny and its ability to reach for greatness. Alexis de Tocqueville first cited it in his study of the United States written from 1835 to 1845, *Democracy in America*.

President Obama played down American exceptionalism in his opening apology tour across Europe after he was elected president. He told the world, "I believe in American exceptionalism, just as I suspect that the Brits believe in British exceptionalism and the Greeks believe in Greek exceptionalism." The difference was that America really is exceptional, and Mr. de Toqueville knew that. As does President Trump, as do I.

The immigration issue will be the most bitterly fought and overly fraught policy front of the Trump Century. Whether or not President Trump can make more fundamental changes, change will come. A vigilant and strong immigration policy is critical to national security and economic policy. All three elements rely on strong, well-regulated borders. Rather than debate these points, the president's foes said they reflect a fear of foreigners and bigotry.

In the primary season for the next presidential election, all of the almost two dozen candidates for the Democratic nomination saw racism and hate lurking everywhere in the United States: in the economy, hiring, compensation, health care, insurance coverage, education, and justice. Systemic racism was endemic to our nation's history and remains firmly in place, they argued.

As President Trump campaigned for reelection, that Dem strategy gained momentum in mid-2019. On July 27, a longtime advisor to the Clinton camp, Joe Lockhart, said this on Twitter:

Anyone who supports a racist or a racist strategy is a racist themselves. 2020 is a moment or [*sic*: of] reckoning for America. Vote for @realDonaldTrump and you are a racist. Don't hide it like a coward. Wear that racist badge proudly and see how it feels.

In September 2019, that sentiment was seconded in the liberal Daily Beast. The banner headline was "Trump Is a Racist. If You Still Support Him, So Are You."

To explicitly deny that you are a racist further entangles you in the wrong debate—on whether that charge is true—and distracts from the real issues. It gives the media more opportunities to mention your name in the tussle over how racist you may or may not be. God forbid that one straight, older, white male should have the temerity to defend another. We had our time on this earth, just like the dinosaurs; we'd best go quietly, with a whimper. Okay, boomer?

The difference this time around was that their target, Donald Trump, refuses to take anyone's guff, to use a polite and old-fashioned term. He may be the closest thing we have today to the John Wayne of politics: a swaggering, combative, proud American willing to be the louder one when confronted by any adversary who crosses a line. His line. This swagger influences the administration from the cabinet on down.

For two decades or more, the Dems' demonization fractured the debate on immigration reform and the need to fix unfair trade. They have kept the most urgent issues in American life off the table. Until President Trump.

CHAPTER TWELVE

FED FIXATION

President Trump had an abundance of visible and vehement opponents to his America First agenda, but one of the most formidable foes confronted him from inside the federal government: the Federal Reserve System. As he pushed hard to reignite growth in the US economy with a bold lineup of tax cuts, business deregulation, and a trade war on China, the Fed was working to undercut everything he was doing.

The Fed had let the federal funds rate, the interest rate banks pay to borrow money from each other overnight, sit at 0 percent for almost the entire eight years of the Obama presidency. But in the first two years of the Trump regime, the Fed would raise it a startling eight times in a row. It would do so for no apparent reason, amid zero signs of rising inflation. That suffocated growth and killed animal spirits—which was precisely the Fed's objective.

The Fed raised the federal funds rate fivefold in twenty-five months, from 0.5 percent in November 2016 to 2.5 percent in December 2018. If the fed funds rate rises, other interest rates also rise, from small-business loans and home mortgages to credit cards. The Fed's hawkish rate hikes vastly increased the marginal cost of borrowing new capital at the whip end of the economy, in new loans and new investment for building new business. That blunted economic growth.

The Fed's unwarranted and poorly timed policies had the same effect as if it were plotting to undermine the agenda of the newly elected president of the United States. For President Trump, that must have been infuriating and inexplicable. In mid-2018, he embarked on a pitched and protracted campaign aimed at jawboning the Fed into halting its relentless rate hikes and start lowering. It was the first time in thirty years that any president had called out the independent institution with a stranglehold on the US economy.

President Trump did it with alacrity. In the next six months, he would send seventy publicly reported messages to the Fed via interviews in the Oval Office, comments at rallies and press briefings, and exclusives on a succession of days with Fox Business, the *Wall Street Journal*, Reuters, Bloomberg, CNBC, and more. And tweets, lots of tweets.

He would prove to be right—the Fed's incessant rate hikes were derailing the upturn in business investment and job growth that the Trump agenda had been creating with great success. That was of less importance to the pixel pundits and indignant economists who rose up to bash President Trump for bashing the Fed. They were aided, as always, by a media ecosystem that covered only the uproar rather than the underlying issues.

All of them were too busy haranguing the president for his caustic style and for daring to tread so publicly on the autonomy and credibility of the Fed. The racism angle so integral to Democratic opposition was unlikely to emerge in the arcane skirmish, so the dictator/authoritarian attack would be the go-to counterargument. By late 2018, they were in high dudgeon. Cue angry-face emojis across the mediascape.

A December 17, 2018, commentary on CNBC.com by a Villanova School of Business economics professor was headlined "Nixon All Over Again: Trump's Fed-Bashing and Interest-Rate Panic Will Cause a Recession, Not Prevent One." A June 24, 2019, *New York Times* article stated, "His grievances echo those voiced by presidents in the 1960s and '70s, though his favored delivery channel—a social media account with more than 60 million followers—is far more public." On July 31, 2019, a former CNN correspondent opined, "Trump is following the playbook of other authoritarian populists: embracing nationalist rhetoric and policies, developing an us-vs.-them narrative ahead of the elections and undercutting the independence of the Central Bank." A *Politico* headline on September 11, 2019, read, "Trashing Fed 'Boneheads,' Trump Calls for Central Bank to Cut Interest Rates to 'ZERO.'" In that story, David Kotok, a frequent commentator on CNBC and the chief investment officer at Cumberland Advisors, declared, "Trumpanomics of Fed bashing and trade war are an economic menace to the United States."

A *Forbes* column on January 16, 2020, was titled "Trump's Simmering Ire at Federal Reserve Chair Jerome Powell Defies Economic Logic." On September 20, 2019, Alan Blinder, a former Fed vice chairman, provided a historical, if unkind, perspective on CNBC.com under the headline: "'I Don't Like Trump,' but Presidential Fed Bashing Nothing New, Says Ex–Fed Vice Chair Blinder."

None of the coverage addressed an insightful point made by the president: that the United States is the safest issuer of government bonds on the planet. Why should we pay *higher* interest rates to investors than those paid by governments in Europe that are wobblier than ours in terms of the chance that they may default on paying them back?

President Trump had pushed his tax plan past Congress, the stock market was on a happy tear, consumers were feeling buoyant, and the jobless rate was falling. Yet a bunch of government bureaucrat bankers in Washington were tripling or quadrupling the underlying cost of borrowing money that could help expand business and drive still more growth—and for no apparent reason at all, other than this one: because they could.

That was an unforgivable policy failure, yet the Fed never admits to a blown call. An old expression says military generals always fight the last war. Add that line to the old joke about the Fed—"often wrong and never in doubt"—and it describes the Fed of the Trump era.

The Federal Reserve is run by people who are sufficiently opaque and wizened to think that they can deliver stable markets and high employment as if they could conjure them up in their monthly meetings. They soon learn that they can't do that. Yet they love trying to do so.

By one count, in a previous run, the Fed had gotten it wrong on nine rate moves in a row. Clearly, it got it wrong eight times in a row in the first two years of Trump. Its paranoiac fears of inflation preempted all other judgment—and not even an actual rise in prices, just a rise in consumers' *anticipation* that prices might rise soon.

That would have been less intolerable if the Fed had had more humility, if it had been consistently modest in its actions and quiet in its pronouncements. The markets would have been a better place. Instead, we saw a Fed pose of stentorian gravitas as it hewed to a policy mired in inertia or simple indifference.

When Donald Trump arrived in Washington, he possessed special wisdom from forty years as a real estate developer in New York, gaining insight into the credit markets, capital investment, the animal spirits of capitalism, and the way the rising cost of borrowing squelches a business's plans to invest in expansion.

Fed officials lack that kind of experience and viewpoint. They are economists, academics, wonks, and former investment bankers, steeped in data and statistics and economic models, when the economy is composed of people and emotions and hope and risk. Data are unable to capture this entirely, much less model it and predict it.

Yet the Fed folks are making political decisions beyond their ken. In this particular saga, they thought they knew as much as President Trump did. They made it clear within the first six months of his presidency that he understood money, credit, and debt far better than they did, both as an institution and as the Federal Open Market Committee.

All of this is plainly true, but for some reason criticizing the high priests of the Federal Reserve was out of bounds. Traditionally, the relationship between the US central bank and the president of the United States was a respectful and restrained one, in public and on the surface.

They need each other. The president appoints the Federal Reserve chairman to a four-year term, but legally, the chairman doesn't serve at the pleasure of the president. For once, Donald Trump, the former star of *The Apprentice*, is unable to invoke his TV catchphrase, "You're fired!"

The Fed chairman runs an independent entity with 22,000 employees, an annual budget of more than $5 billion, the Federal Reserve System of a dozen regional Federal Reserve Banks, the seven-member Board of Governors, and the twelve-member Federal Open Market Committee, made up of the seven governors and the New York Federal Reserve Bank head, plus four regional bank chiefs.

The Fed has a stranglehold on the two choke points—interest rates and the money supply—that can stall or stoke the economic growth that every president needs for reelection. Established in 1913, the Fed today is guided by two mandates: price stability (i.e., tamping down inflation expectations on the part of consumers, businesses, and the markets) and maximum employment (i.e., a low jobless rate).

These are opposing forces. If interest rates are too high and the money supply is tight, businesses will forgo borrowing to expand and job growth will stall. If job growth spurts up past the Fed's uninspired expectations, it might spark inflation fears in the economy and, more important, at the Fed itself. That possibility might prompt the Fed to raise rates sooner to blunt the enthusiasm. But if it raises them too much, it kills the party. It is a delicate balance. Donald Trump doesn't do delicate.

In addition to setting the base rate for other interest rates (choke point number one), the central bank controls the size and flow of the money supply (choke point number two). When the financial markets panic, the Fed floods the system with extra liquidity by buying government and private bonds from Wall Street investment houses. After the Wuhan pandemic, its balance sheet was likely to swell up to $7 trillion or $8 trillion in assets.

When the Fed raises the federal funds rate, it can trigger rate increases most everywhere else in the economy. After six months of increases in the first half of 2018, the rate on a thirty-year fixed-rate mortgage had risen by 0.6 percentage points, from 3.95 percent to 4.54 percent. The *Wall Street Journal* said that would add $100 to the monthly mortgage payment on an average-priced home. The four hikes the Fed imposed in 2018 would cost households an extra $100 billion in higher mortgage costs annually, the personal finance website MagnifyMoney found.

Fed officials started fretting about the Trump economic agenda even before the transition of power in January 2017. Let other critics dismiss the Trump blueprint of tax cuts and deregulation as fanciful or undoable; it posed one big inflation risk in the view of the Fed: Trump was proclaiming a target of 4 percent annual growth, double what the forecasters at the Fed itself were saying was plausible in the long term.

The Trump campaign's promises made the Fed officials more fretful that a surprise jump in growth could ignite inflation expectations. That might send the economy into a destructive manic and depressive cycle.

President Obama took office in January 2009 in the midst of the global financial collapse and the Great Recession, which ended at the end of June 2009. The federal funds rate stayed at 0 to 0.25 percent for most of the eight years of his presidency. Growth stayed that slack. The Fed had raised rates only once, at the end of 2015, to 0.5 percent, by the time the 2016 election occurred in November.

The Fed chairman at the time was Janet Yellen, appointed in February 2014 by President Obama. Candidate Trump had already started bashing the Fed in the 2016 campaign. He had made it clear that if he won the election, he would replace her when her term ended in early 2018. That would break a forty-year tradition of reappointing the prior president's Fed chairman. In May 2016, he had told CNBC, "She is not a Republican. When her time is up, I would most likely replace her because of the fact that I think it would be appropriate."

In a campaign ad, he labeled Yellen among the "global special interests" who had ruined life for middle America. In September 2016, by then the Republican nominee, he criticized Yellen on CNBC, this time for the Fed's allowing the US dollar to grow too strongly against foreign currencies. He also said she should be "ashamed of herself" and that central bankers were "very political" for keeping interest rates so low to benefit the Obama administration, "so Obama goes out and let the new guy . . . raise interest rates . . . and watch what happens in the stock market." Those remarks were on target, and they rankled the pundits and the media.

Two months after Trump gave that interview to CNBC, he won the election. A few weeks later, on December 15, 2016, right before he took office, the Federal Reserve raised interest rates a quarter point, to 0.75 percent. It was the second rate hike in eight years, and Fed chairman Janet Yellen had a full year left in her term.

The new hike came in a year in which GDP growth finished at a listless 1.6 percent for the entire year. It was inexplicable.

Under her watch, the Fed raised rates three more times in a row, in quarter-point increments, up to 1.5 percent by year-end 2017. In late October, a week before he was about to name a new appointee to be chairman of the Federal Reserve, President Trump appeared on my show on Fox Business to discuss the matter.

On air, the president asked me point-blank, "Do you have a preference, out of curiosity? Tell me who your preference is. I would love to hear it. I only want that from people I respect." He can flatter to a fault. I told him, "I personally believe that Janet Yellen might be worth keeping."

President Trump responded that she "was very impressive" in their recent meeting in the Oval Office, adding "I like her a lot." "I mean, it's somebody that I am thinking about. I would certainly think about it."

A week later, on November 2, 2017, President Trump nominated "with great pleasure and honor" Jerome Powell as Yellen's replacement to be chairman of the Fed. He was hoping for a lot more cooperation with—and accommodation by—a chairman of his own choosing. Powell had been serving on the Federal Reserve Board of Governors since 2012.

At a Rose Garden ceremony marking the nomination, President Trump told the media he had picked Powell for the job because "He's strong, he's committed, he's smart." Powell told reporters that he was "committed to making decisions with objectivity, based on the best available evidence, in the long-standing tradition of monetary policy independence." If he could have put "independence" into italics, he would have done so.

At the gathering, President Trump went out of his way to thank Janet Yellen, "a wonderful woman who's done a terrific job. We have been working together for 10 months, and she is absolutely a spectacular person. Janet, thank you very much. We appreciate it." He said she had served "with dedication and devotion" and that the Fed is "respected all around the world and is crucial to our economic prosperity."

Whether the president's praise was gracious or insincere—take your pick—Yellen was unforgiving. In February 2019, she expressed doubt that President Trump understood economic policy or the mission of the Fed: "I doubt that he would even be able to say that the Fed's goals are maximum employment and price stability," she told NPR. She expressed concern about his criticism of the Fed and her successor.

In December 2018, a month after President Trump nominated Jerome Powell as Fed chairman, the Senate Committee on Banking, Housing, and Urban Affairs voted 22–1 in favor of sending the nomination to the US Senate for final approval. The sole dissenter was Senator Elizabeth Warren, who had clashed with Powell over regulation in confirmation hearings.

Powell assumed the chairman's post on February 5, 2018. At the very next meeting of the Federal Reserve Board of Directors, in March, the Fed hiked the interest rate again—and signaled that it would keep hiking rates. GDP growth had finished 2016 at a subpar 1.6 percent, and it had rebounded to 2.4 percent in 2017, despite four fed funds rate hikes in a row that year. That marked a *50 percent faster* economic growth rate in just one year.

Now, as 2018 approached, the Fed found the strength of the Trump economy unnerving. The minutes of the Federal Open Market Committee's March meeting revealed that Fed officials were wary of the tax cuts Congress had passed at the end of 2017. They were uncertain what effects the cuts would have, "partly because there have been few historical examples of expansionary fiscal policy being implemented when the economy was operating at a high level of resource utilization."

The Fed was nervous that the tax cuts would spur inflation rather than real growth. Forecasts now showed that the Fed expected to raise rates two more times in 2018, and some economists were looking for three. That was rather perplexing to me, as the economy showed zero signs of inflation, the Fed's number one worry.

The rate hike of March 22, 2018, was particularly ill timed. On the same day, the Trump administration would take another step toward escalating its confrontation with China on trade, releasing its report detailing China's many unfair trade practices related to technology transfer, intellectual property, and innovation.

Just three weeks earlier, on March 1, President Trump had slapped China by setting a 25 percent tariff on all imports of steel and a 10 percent tariff on all imported aluminum. China accounts for 50 percent of the worldwide steel production and 67 percent of global output of aluminum.

The combination of Powell's surprising hawkishness and the Trump tariffs roiled financial markets. It was hard to put the two together. Fed officials had warned that a trade fight could be a drag on the economy. That would call for the Fed to step away from raising rates—at least long enough to see if those concerns were valid.

By raising rates a month into the new chairman's term and signaling more hikes ahead, the Fed risked weakening the US economy just as we were confronting China. It struck many as troubling. The Fed seemed to be raising rates in lockstep without considering the risks to the economy, even as we were launching a trade war.

The Trump administration's pressure campaign on the Fed began unofficially on April 2, 2018, in an interview on CNBC between anchor Kelly Evans and Peter Navarro, then the director of the White House National Trade Council. Navarro, a deft and seasoned pundit in debates on cable news, was Trump's most voluble hawk on China, trade, and the Trump tariffs.

On the last question, Evans got arcane: The economy, she said, is great, but why is the yield on the benchmark ten-year Treasury note down a bit today? It had fallen from almost 3 percent to 2.7 percent. Falling ten-year yields can show that investors are worrying about long-term growth. She told Navarro, "Frankly, no one around here can really explain it. What do you think is going on?"

Navarro redirected that into a tacit swipe at the Fed. Evans instantly caught the gravity of what Navarro was saying. She went in for the pin.

NAVARRO: Yeah, I was a little puzzled when the Fed announced three rate hikes before the end of the year, because when I look at the chessboard, I don't see any inflation to speak of in the economy. And a lot of reasons are things like the president's tax cuts, which are going to stimulate a lot of investment, productivity, growth, downward pressure on wages, and, remember, the supply side effects of deregulation. . . .

EVANS: Yeah—but does that mean, Peter—which is great. You're right and that's the ideal outcome. But final question here, does that mean that there's no reason to keep hiking interest rates?

NAVARRO: Well, that's the Federal Reserve job. All I can tell you is I was surprised when I saw that announcement based on my read of the inflation chessboard. I remember in the late '90s when Alan Greenspan was hurrying to raise interest rates and we had tremendous productivity growth and low inflationary pressures then. So, again, you know, bottom line here is that the market—

EVANS: Is it a mistake? Was it a mistake then? Is it a mistake now?

NAVARRO: I won't say that. I'm not in—that's the Federal Reserve chairman's lane. And we've got a good—Kevin Hassett, the Council of Economic Advisers.

EVANS: I think you're hinting that it's—hinting at a mistake.

NAVARRO: All I'm saying is to the investors watching CNBC here is that the economy looks very, very strong on all parameters and doesn't appear to be significant inflationary pressures that would detract from that strength, so it's all good.

EVANS: All right Peter Navarro. Thank you for your time.

NAVARRO: My pleasure.

Translated from Fed-speak: Yes, Navarro was saying that the Fed made a mistake.

It was mild stuff, but it provoked gasps in the corporate financial media. The end was nigh: a Trump administration official had dared to publicly voice skepticism about the Fed's monetary policy! Contradictorily, some of the loudest objections to Navarro's remarks came from the people who had argued the China tariffs would be a drag on the US economy; they should have agreed with Navarro that the Fed had raised rates too high.

For all the waves of disapproval in Fed circles, no one at the Central Bank was intimidated. At its next regular meeting, in June 2018, the Board of Governors voted unanimously to raise the federal funds rate target by another quarter percentage point, to a range of 1.75 percent to 2 percent. Instead of hiking the rate three times in 2018, they were now projecting four increases.

As if to say, "Tweet that, President Trump!"

Eight of the fifteen board members indicated that they expected at least four rate increases in 2018, up from seven votes in March and just four votes in December. Further, most Fed officials said that the rate would go up at least three more times in 2019 and at least once more in 2020. That would drive the rate up to an upper range of 3.5 percent by the end of President Trump's first term—and strangle any attempt at the 4 percent economic growth he had so publicly promoted.

For the first time in four years, the Fed's official policy statement omitted language that said officials expected to hold their target rate below the neutral level "for some time." In other words, the age of easy money and superlow interest rates was coming to an end. Monetary policy accommodation would be over soon.

Dropping those three simple words, "for some time," was widely perceived as a hawkish move by the Fed—a signal that rates were heading higher. It was not even a question anymore. Now the Fed said only that the "timing and size" of future rate increases would be determined by many factors. They were now inevitable, in any case.

Powell had altered the stance of monetary policy. He also had changed, literally, the stance of the Fed chairman. Ben Bernanke and Janet Yellen had both conducted press conferences seated behind a desk, a pose perhaps inspired by their academic backgrounds and befitting a government official appointed to carry out bureaucratic or clerical duties.

Powell chucked the desk and stood, instead, at a lectern, as a president does when addressing the press. He declared that he would speak straight to the American people in plain English. He announced that, instead of holding press conferences after every second meeting, as Yellen had done, he would hold them after every meeting. He was transforming the role of the Fed chairman from a behind-the-scenes post to one of public leadership.

Powell's comments at the press conference after the June meeting underscored that notion: "As the economy has strengthened and as we've gradually raised interest rates, the question comes into view of, how much longer will you need to be accommodative and how will you know?"

The Fed had *already* stopped being accommodating. It had just raised rates for the fifth time in less than two years. Powell mentioned that the Fed was being "very careful not to tighten too quickly." He added, in a message aimed at 1600 Pennsylvania Avenue, "We had a lot of encouragement to go much faster, and I'm really glad we didn't."

Translation: But we could do so at any point, so avoid messing with us.

Chairman Powell was making the same mistake I have seen a procession of Fed chairmen make over the decades. When you have a hammer, suddenly everything begins to look like a nail. As soon as they start their new term, they feel compelled to test themselves by raising rates to make their claim on new ground.

Even a sage high priest of Fed black arts made the same error. Alan Greenspan, who served five terms as Fed chairman from August 1987 to January 31, 2006, did so at the start of his tenure. Peter Navarro alluded to it in his interview with Kelly Evans on CNBC.

The Greenspan rate hikes helped send the economy into a deep downturn after the market crash on Black Monday, October 19, 1987.

Powell had convinced himself he could read the markets and economic signals, when his background lent itself to other strengths. He had started his career in New York as a lawyer and investment banker, then served Bush 41 as undersecretary of the Treasury, after which he had been a partner for almost a decade at Carlyle Group, a giant private equity fund with $200 billion in assets. His perspective was that of an investment banker. President Trump approached it from the perspective of a builder.

President Trump took Chairman Powell's higher public profile not so much as a provocation as an invitation. If Powell could speak in plain English from the bully pulpit of his new lectern, surely Trump could publicly address Fed policy.

Powell had recently appeared on Capitol Hill for the semiannual ritual in which the Fed chair testifies before Senate and House panels, where he had been grilled by lawmakers about why wage growth was so low, despite the low level of unemployment. Democrats demanded to know what the Fed would do to accelerate wage growth. There was little concern about inflation or interest rates at the hearing.

None of the senators held back in homage to tradition and Fed independence. Witnessing the spectacle, President Trump must have wondered: Am I the only person in the world who is barred from questioning the Fed's monetary policy?

This highlights how strange the conventions of Washington had become regarding the Federal Reserve. Senators and congressmen were permitted—even expected—to pepper the Fed chairman with questions and challenges. But the president of the United States was expected to keep his mouth shut, no matter how egregious the mistakes of the central bankers became—even though he lacked the authority to fire the Fed head.

But President Trump was hardly going to keep quiet. The mystery is why anyone had ever thought he would.

The June 2018 increase by the Fed convinced both economists and the markets that it was hell bent on raising rates and shrinking its balance sheet. Michael Gapen, the chief US economist at Barclays Investment Bank, told the *Wall Street Journal* that Powell's press conference and the Fed's statement had implied that the rate-hiking policy was actually on autopilot.

Talk of a looming recession filled the media. Three days after the June hike, the *Journal* reported that one forecasting model, from BBVA bank in Spain, now predicted that the chance of a recession had more than tripled in six months, from only a 5 percent chance in the next twelve months to a 16 percent chance. It was a big jump.

Counting the first rate hike in December 2016, a month before Trump had taken office, up until that point in June 2018, the Fed had raised the rate six times, starting at 0.5 percent before the first increase and raising to 2.0 percent. That may seem small; in fact, it amounted to more than tripling the underlying cost of borrowing funds. It was a shocking rise, and the Fed overdid it terribly. It would go on to raise the rate two more times, pushing it up to 2.5 percent in the ensuing six months to December 2018.

Frustrated, President Trump in July 2018 started a six-month pressure campaign, criticizing the Fed in the most direct public assault ever launched by any president. In the next six months, he would devote some seventy publicly reported messages aimed at addressing the Fed.

He started on July 19, 2018, in an interview at the White House with CNBC: "I'm not thrilled. . . . I don't like all of this work that we're putting into the economy and then I see rates going up." That was enough to prompt CNBC.com to call it a "stinging and historically rare criticism." The *Journal* said that the president's remarks "break with tradition that presidents refrain from commenting on monetary policy." Former Dallas Federal Reserve Bank president Richard Fisher warned that the president was overstepping his authority. Former Federal Reserve governor Frederic Mishkin told CNBC that the Trump sniping was "not particularly welcome" and a "danger sign."

Obama's Treasury secretary Lawrence Summers told reporters, "Likely result of Presidential intervention is higher rates as Fed needs to assert its independence."

The next day, President Trump tweeted on the Fed, complaining that China and the European Union were manipulating their currencies and interest rates to reduce them, to our disadvantage. Four weeks later, he made private comments to wealthy donors at a fund-raiser on Long Island, telling them he had expected "Jay" Powell to be a cheap-money Fed chairman and lamenting that he had raised rates, instead.

Ten days later, in an interview in the Oval Office with Bloomberg News, he said, "We are not being accommodated. . . . I don't like that." He added, in a moment of diplomacy, "That being said, I'm not sure the currency should be controlled by a politician." He told the Bloomberg reporters that he didn't regret appointing Powell.

A month later, on September 26, the Fed raised the federal funds rate *again*—up another quarter point, to 2.25 percent. President Trump told reporters at a press conference in New York, "Unfortunately they just raised rates a little bit because we are doing so well. I'm not happy about that."

At an October 10 rally in Pennsylvania, he remarked, "They are so tight. I think the Fed has gone crazy." Later that day, he said in an on-air call with Fox News that the central bank was "going loco" by raising rates. Six days later, he told Fox Business Network that the Fed is the "biggest threat" to the economy. He added that the central bank was "independent, so I don't speak to them, but I'm not happy with what he's doing because it's going too fast."

Then, on October 23, the president told the *Journal* that "maybe" he regretted appointing Powell as Fed chairman, although he added, "I'm not going to fire him." With disarming transparency, he said he was sending a message to Powell, while acknowledging that the Fed is independent.

With the next Fed meeting on December 19 approaching, President Trump stepped up the pressure campaign, carpet-bombing the issue six times in the next four weeks. He broached the topic on November 20 at a press briefing.

On November 26, he did an interview with the *Journal*, telling the paper "I think the Fed right now is a much bigger problem than China. I think it's—I think it's incorrect, what they're doing. I don't like what they're doing . . . but the Fed is not helping." A day later, he told the *Washington Post* that he was "not even a little bit happy with my selection of Jay. . . . I think the Fed is a much bigger problem than China." He added, "They're making a mistake because I have a gut, and my gut tells me more sometimes than anybody else's brain can ever tell me."

That was followed by an interview with Reuters on December 11, one week before the Fed board was to meet and vote on whether to raise rates again, in which he said, "I think that would be foolish. . . . I need accommodation." He called Chairman Powell a "good man." "I think he's trying to do what he thinks is best. I disagree with him. I think he's being too aggressive, far too aggressive, actually far too aggressive."

A week later, the president put up a last, plaintive tweet, saying it was "incredible" that the Fed is "even considering yet another interest rate hike. Take the Victory!"

Two days later, on December 19, the Fed raised rates one more time, up another quarter point to 2.5 percent—five times as high as the rate had been in November 2016 when Trump had been elected.

President Trump's agenda had sparked GDP growth in spite of the Fed's constant pushback, making this feat all the more impressive.

All that public lobbying had failed to move anyone at the Fed. President Trump was furious. On December 21, Bloomberg reported that President Trump had discussed firing the Fed chairman, which would violate the law, undermine the Fed's autonomy, and rattle the markets. The Trump administration spent the next few days softening the message. On December 23, Mick Mulvaney, Trump's pick as his next chief of staff, told reporters he had spoken with Treasury secretary Steven Mnuchin and had learned that the president "now realizes he does not have the ability" to fire Powell.

Then, on Christmas Eve 2018, President Trump tweet-bashed the Fed again:

> The only problem our economy has is the Fed. They don't have a feel for the Market, they don't understand necessary Trade Wars or Strong Dollars or even Democrat Shutdowns over Borders. The Fed is like a powerful golfer who can't score because he has no touch—he can't putt!

Golf trash talking; *now* the president was getting serious.

On Christmas Day at the White House, reporters asked President Trump about Powell. He answered, "Well, we'll see. They're raising interest rates too fast. That's my opinion. . . . They're raising interest rates too fast because they think the economy is so good. But I think that they will get it pretty soon."

The next Fed meeting for a vote on interest rates came on January 30, 2019. For the first time in two years, the Federal Reserve Board of Governors voted to keep interest rates at their current levels. At long last, a pause.

Five days later, on February 4, the president hosted Fed chairman

Jerome Powell, the Fed vice chairman, and the Treasury secretary at a casual dinner at the White House. Immediately afterward, the Fed released a statement saying that its representatives hadn't discussed Powell's expectations for monetary policy. They hadn't had to discuss the matter; the president had made his views known.

After months of Fed fury and President Trump's barrage of persuasion and intimidation, the Fed had finally stopped raising rates. After eight hikes in a row over two years, it stood back. It looked like a victory for President Trump. It wasn't.

In the next month or two, President Trump would place two calls to Chairman Powell. Though the Fed's hawkish stance on pushing rates higher seemed to be softening, he would stay on it, keeping up the pressure on the Fed to start cutting rates. Yet the Fed let the rate stand at 2.5 percent and refused to cut it for the next seven months.

That stubborn stance by the Fed whacked the economic growth the Trump administration was endeavoring so mightily to achieve. The brain trust at the central bank was supposed to tap on the brakes of the economy and, instead, had jammed down too hard and sent growth skidding. The numbers show it.

- In 2016, GDP growth was a torpid 1.6 percent.
- In 2017, Trump Year 1, growth jumped to 2.4 percent, despite four rate raises.
- In 2018, Trump Year 2, growth jumped to 2.9 percent, despite four *more* hikes.
- In 2019, Trump Year 3, growth slowed sharply to 2.1 percent. The Fed boosts had killed the Trump rally—as intended.

A 2.5 percent federal funds rate for banks' overnight loans seems small. But it was vastly higher than in Germany and elsewhere, places that were a higher credit risk than the United States and that therefore should have been paying higher interest rates on their government bonds. Instead, the strongest among them, led by Germany, were *charging* investors a fee to put their money into zero-rate bonds, resulting in "reverse" interest rates.

Another drag on growth, in the president's view, was that the Fed's policies were letting the US dollar rise too high against the value of currencies in Japan, China, and Europe. A strong dollar and how much it can buy in goods is an indication of the country's underlying strength, just as a stock price indicates the same of a company. If the dollar gets too strong, it makes the cost of US exports too high for buyers in other countries; they can get the things cheaper elsewhere in a local currency.

None of that swayed the Fed's resolute and devout devotion to ensuring the highest possible interest rates without strangling economic activity. Then suddenly the Fed began *cutting* rates to offset the strangling effects of its own increases. On August 1, 2019, it cut a quarter point to 2.25 percent; just six weeks later, on September 19, another quarter-point cut to 2.0 percent; on October 31, a third cut to 1.75 percent. Three cuts in three months—a clear sign that someone at the Fed had previously erred egregiously.

Without the Fed's deigning to admit it, it was tacit proof that it had gotten it wrong and raised rates way too high and left them there for far too long. President Trump had been right all along about the Fed's raising rates too high when it should have been cutting them without fear of inflation—the only reason to push rates up.

Events had overtaken any fury over Trump's having the temerity to criticize Fed policy. As the Wuhan pandemic began to threaten worldwide commerce in March 2020, the Fed imposed emergency cuts: on March 3 a half-point cut, rather than the usual quarter point, down to 1.25 percent; and then a full point on March 15, down to just 0.25 percent, a crisis level.

Whoever leads the United States after the 2020 election, the prospects for future clashes with the Fed will loom large. The next president will need all the help the Fed can offer, yet the Fed will be looking to tamp down the comeback at the first sign of giddiness, and only the Fed gets to decide when that is.

Economists find this acceptable. Before Trump assumed office, an economist for the Japanese securities firm Nomura told the *New York Times* that the real effect of the Trump agenda, if it stokes economic growth, will be to prompt the Fed to raise rates to tamp down the new growth before it even gets under way. He explained, "If we have a big stimulus, the logical thing for the Fed to do is to raise rates faster. There isn't a whole heck of a lot of scope to just let the economy run under those circumstances. There's a question about whether fiscal stimulus under Trump just leads to higher interest rates."

So even though the American people elected Donald Trump, it was really up to the unelected Federal Reserve Board of Governors to decree whether his policies would be permitted to push the accelerator on growth. President Trump's target of 4 percent growth, twice what the experts said was possible, was beside the point. The Fed was inclined to overrule the people, and there was nothing anyone could do about it.

This is perverse. This is a national problem. The Fed answers to no one, really. It certainly doesn't answer to the president; otherwise it would have stopped hiking rates much earlier. It isn't elected, duly or otherwise, by the American people, yet it holds the power to deny economic success to a newly elected president.

There was no telling how the Federal Reserve would proceed as the rebuilding of the United States began. For President Trump, one concern had to be: What if these guys blow it again? For decades the Fed has been autonomous and off-limits to protect it from political interference. Yet the central bank is removed from the people whose lives can be made significantly better—or worse—as a result of its good and bad calls.

The Fed's two mandates are disconnected from the things that help Americans thrive economically and build better lives. The Fed is obsessed with muzzling inflation expectations, which requires suppressing growth. It must maximize jobs, which requires *better* growth. It wants a strong dollar, which hurts US imports—and impedes GDP growth. The Fed is risk averse and more intent on avoiding wild swings than delivering an economic rebound that could help millions of Americans find better-paying work. Workers and business owners, by contrast, want vibrant growth, strong wages, loose credit, and easy borrowing. These are contradictory agendas.

When the Fed makes a terrible call, as it did in the first two years of the Trump administration, it answers to no one. President Trump answers to the American people. The disconnect should be addressed, to strengthen the Fed's accountability to the people it is supposed to serve. I doubt that any of the people at the Federal Reserve feel their purpose is to serve the people. They serve the banks and trading partners and global elites.

Trump's views were right, and they went largely ignored by the corporate media and even by economists who knew he was correct. It was easier to dismiss and depict Trump's tweets as Fed bashing than to engage seriously on the economic analysis beneath it. It was the same old pattern: assume Trump was shooting from the hip, and criticize him for his impulsiveness without taking a deeper look at what he had said and what he hoped to accomplish.

In the Trump Century, for the president to restrain his comments and avoid pointing out the Fed's flawed path would disserve the American people—attention must be paid, something had to be said. Who else better to say it?

After all that fracas and back-and-forth, and in the aftershocks of the Wuhan pandemic, the Fed would cut the interest rate to zero. It looks likely to stay there for several years, as the United States rebuilds from the months-long shutdown of commerce and regular life. It was the worst possible way to get there, but it will help whoever is elected president rebuild the US economy better and smarter and bigger than ever before.

CHAPTER THIRTEEN

THE NEW OLIGARCHS

S ilicon Valley has gone all in on hating Donald Trump.

That is true of the rank and file, filled as it is with superlefty millennials steeped in the anti-American, antibusiness indoctrination they learned at some of the priciest universities in the country: Harvard, Yale, Princeton, Berkeley, Stanford. It is especially true among the New Oligarchs of Silicon Valley, the billionaires and their consigliere at the new-generation titans that dominate high tech.

The goliaths include the five collectively known on Wall Street as the FAANGs: Facebook, Amazon, Apple, Netflix, and Google. Throw in Twitter as the junior member. Together they are worth $5 trillion in market value—equivalent to the GDP of Japan, which has the third largest economy in the world. Combined, they reach more than 2 billion people around the world. Eighty percent of the people on the internet use one or more of these websites.

Their forebears who dominated Silicon Valley for decades were engineers and venture investors from the baby-boom generation. Their wealth was accumulated gradually after ten or twenty years of toil. They sold hardware and software in predictable upgrade cycles: Hewlett-Packard, Intel, the old Fairchild Semiconductor, Sun Microsystems, Oracle, Microsoft, parts of IBM and Xerox, and others of the old guard. They were more conservative than today's tech stars, and most certainly, they were quieter about the political views they held. They courted corporate enterprises and happily sold to everyone; why risk offending half your prospects by picking a side in a political debate?

The New Oligarchs are younger and more flamboyant. They are far more liberal, and louder about it. They are also far richer, with nine or ten digits to their net worth (or even twelve, in the case of Amazon founder Jeff Bezos, at north of $140 billion). It came more rapidly than in the past. They rely less on selling to the executives running corporations and more on competing for the attention and time of hundreds of millions of consumers. They buy media companies and raise millions of dollars in campaign contributions for Democrats.

They also lobby for the same old trade agenda that ripped off American workers for decades and gave the American people cheap flat-screen TVs from China.

The concern is that, unlike their predecessors, this new wave reaches an audience of hundreds of millions of people and provides them with content—entertainment, news, government statements, political views, information and misinformation, and worse. They possess the power to shape our world and how we see it and how we feel about it. They can lead us to feeling elated or feeling sadness.

They can get us to share outrage or cute kitten videos. We can see a stream of headlines about how well the United States is rebounding from the Wuhan pandemic or a different stream about all the death and despair.

The triumvirate of Google, Facebook, and Twitter form the

phalanx of a woke movement across the entire tech industry. They have the power to sway a presidential election—and *rig* an election. If Facebook were to send out a "Go vote" reminder only to undecided Democratic voters in four swing states in the Midwest and intentionally leave out all Republicans and right-leaning independents, the social media platform could swing the presidential election in 2020.

This would be of less concern if the big tech platforms were impersonal, monolithic, and neutral, if they strived to be objective, passionless, and nonjudgmental in hosting all comers and welcoming all manner of opinions.

But they take the opposite tack. Facebook, Google, and Twitter are partisan in every way, and they were anti-Trump, virulently and traumatically so, from his campaign through his presidency. The New Oligarchs and their geek legions are ultraliberal and so woke it hurts, like a college freshman in Humanities 101.

This would escalate into a political and legal confrontation in late May 2020, as President Trump ran the country's response to the Wuhan pandemic and the nationwide lockdown. Twitter took it upon itself to initiate a fact-checking program addressing President Trump's tweets—even as it let the president's enemies run rampant on the platform and allege anything they wanted to.

Chinese government officials on Twitter were free to accuse the US military of introducing the Wuhan virus to Wuhan. House co-conspirators in the coup attempt, including the treacherous Adam Schiff, were welcome on Twitter to lie to the American people and sling mud at our president.

But Twitter stepped in like a finger-wagging censor when Donald Trump dared to tweet a message impugning the security and authenticity of a presidential vote by paper mail-in ballots. In California, the Democratic governor and others were trying to use the Wuhan crisis as an excuse to shutter voting precincts and have everyone vote by mail-in—an invitation to rampant fraud. At 8:17 a.m. on May 26, the day after Memorial Day, President Trump tweeted:

There is NO WAY (ZERO!) that Mail-In Ballots will be anything less than substantially fraudulent. Mailboxes will be robbed, ballots will be forged & even illegally printed out and fraudulently signed. The Governor of California is sending Ballots to millions of people, anyone. . . . .

. . . . living in the state, no matter who they are or how they got there, will get one. That will be followed up with professionals telling all of these people, many of whom have never even thought of voting before, how, and for whom, to vote. This will be a Rigged Election. No way!

That was a statement of opinion and concern from the president. Yet Twitter decided that it was a misstatement of facts about paper ballots. It tagged those two tweets with a blue exclamation point followed by a line and a link: "Get the facts about mail-in ballots." Click on the links, and you get stories from the left-wing Fake News: the *Washington Post*, the liberal *The Hill*, and NBC News, the lib sibling of MSNBC.

This condescending and undermining act by Twitter offended President Trump. On May 27, he fired off a litany of threats that were entirely deserved:

Twitter has now shown that everything we have been saying about them (and their other compatriots) is correct. Big action to follow!

Republicans feel that Social Media Platforms totally silence conservative's voices. We will strongly regulate, or close them

down, before we can ever allow this to happen. We saw what
they attempted to do, and failed, in 2016. We can't let a more
sophisticated version of that. . . .

. . . . happen again. Just like we can't let large scale Mail-In Ballots
take root in our Country. It would be a free for all on cheating,
forgery and the theft of Ballots. Whoever cheated the most would
win. Likewise, Social Media. Clean up your act, NOW!!!!

It was an arrogant and foolhardy move by Twitter. At the time, the
state attorneys general in forty-eight states were jointly investigating
Google, Facebook, YouTube, and Twitter for antitrust violations and
other misdeeds, as was the Department of Justice. Attorney General
William Barr had been blunt in his public statements about the social
platforms' stepping way beyond their legally protected roles as neutral platform providers. Now Twitter had united all of them against a
common enemy.

Even more egregiously, by that time Twitter had named a head of
site integrity, an insider who "leads the teams responsible for developing and enforcing Twitter's rules," as the site explained it. That person,
Yoel Roth, was an avowed Trump hater, posting on his own @yoyoel
Twitter feed that Trump supporters had voted for "a racist tangerine,"
that there were "ACTUAL NAZIS IN THE WHITE HOUSE," and
that Trump advisor Kellyanne Conway was akin to Hitler's minister
of propaganda, Joseph Goebbels. Twitter should have fired Yoel Roth
long ago.

The whole thing revealed the hubris and impunity of the New Oligarchs: they didn't care about opposing views, and they knew what
was best for America. Only a year earlier, Twitter CEO Jack Dorsey
had been summoned to the White House to meet with the president,
who had expressed concerns about censorship of the Right. Thus,

Dorsey had imposed the insulting fact-check warning on the Trump tweet even after having met with the president personally, a direct insult that was bound to have consequences.

You do not mess with the president's Twitter following and walk away unscathed. He had been able to use Twitter, primarily, and other social media to get his unfiltered message to millions of supporters. It had provided a major counterbalance to the left-wing corporate media, which had given up all pretense of objectivity and taken up the cause to get Trump.

All in all, the president had almost 200 million accounts following him on the social media platforms by May 2020. That included a combined total of 123 million accounts on Twitter, including 80.3 million for @realDonaldTrump, from which he had issued more than 52,000 tweets; 20.1 million for @POTUS; and 22.5 million for @WhiteHouse; as well as almost 35 million followers on Facebook (29 million for Donald J. Trump, almost 6 million for President Donald J. Trump), almost 20 million on Instagram, and half a million on YouTube. Twitter had dared to intrude on the president's conversation with his people, with no call and no cause for doing so.

Twitter's leaders, like all of the New Oligarchs of Silicon Valley, were so arrogant and so blindly libbed out that they no longer saw themselves as providers of a platform who sat it out while the people jousted for attention and followers; now they saw themselves as high overlords guiding the debate and decreeing what was accurate.

The old Silicon Valley was a meritocracy; the new one is a collection of oligarchies. Their executives and managers form the inner party, their workers are the shock troops spreading their politics, and God help the rest of us.

These companies devoted their vast resources to ending the Trump Century before it could take hold. Their tactics included relentless censorship of Trump supporters, quibbling fact checks from "third-party fact-checkers" with their own biases, and the use of algorithms to cut down on the engagement and sharing of conservative content.

In the 2016 presidential election, Google may have swayed an extra 2.6 million votes in favor of Hillary Clinton by rigging search results, according to Senate testimony in 2019. That would account for 93 percent of the 2.8 million votes by which Hillary Clinton won the popular vote (2,868,691). That swing cost zero for Google to deliver.

Writ large, this would render moot the pitched debate over big-business contributions to political campaigns and attempts to buy influence. Google can do it at no cost at all.

In 2020, the Holy Trinity of wokeness—Google, Facebook, and Twitter, or GooFaTwit for short—could skew search results and news feeds to push upward of *15 million* votes to the Democratic Party. They were likely to rev up their manipulation because in 2016, they had thought Hillary would win—so they had slacked off. Now President Trump held office, and their assault was expected to go full tilt.

Those assertions were made in public testimony to a Senate subcommittee in July 2019 by a psychologist who had spent almost seven years studying the techniques at Google, Facebook, and Twitter and assessing their impact on elections. To the extent that the mainstream media covered his comments at all, none followed up.

The New Oligarchs' ties to the far Left and the Democratic Party run deep. Google has been masterful at serving the Dems and gaining government clout. It deployed its search secrets to help reelect President Obama. In return, Google wielded an unparalleled level of influence over the executive branch. Google's top lobbyist, Johanna Shelton, visited the Obama White House 128 times. Other Google representatives met with White House officials 427 times in eight years, according to the Campaign for Accountability.

The Obama White House and Google created a revolving door for Swamp creatures. In all, 250 people left the White House to work for

Google, or vice versa. That spread the Obama brand of left-wing politics to the already progressive Silicon Valley and threw gasoline onto a raging fire.

After President Obama left office, naming an Obama administration alumnus to the board of a Valley company became as de rigueur as offering vegan options in the campus cafeteria. Valerie Jarrett, his closest advisor, joined the board of Lyft. Obama advisor Daniel Pfeiffer picked up a gig at the crowdfunding site GoFundMe. The infamous Obama "wingman" Attorney General Eric Holder was hired by Airbnb. Former campaign manager David Plouffe has worked for Uber and for Mark Zuckerberg's foundation, the Chan Zuckerberg Initiative.

Former national security advisor Susan Rice was added to the board of Netflix. The streaming giant grabbed the crown jewel—a "development deal" for Barack and Michelle Obama to produce shows and films for it, beating out a rival bid from Apple. Netflix's chief content officer, Ted Sarandos, is a friend of the Obamas. During Obama's first term, Sarandos's wife, Nicole Avant, served as US ambassador to the Bahamas.

On top of all that, there was a $65 million deal for the Obamas' separate memoirs from Penguin Random House, owned by Bertelsmann of Germany. The profit outlook for those deals is debatable, but they are a way of buying the favor of progressive politicians, the liberal media, and the Left at large by paying tribute to their patron saint.

The next time Netflix's children's program lineup seems a little heavy on the liberal, LGBT, socialist agenda, take a look at who is running the operation and the progressive politicians they have hired.

In the 2016 campaign, employees from Google made up the largest employee donor group to the Hillary Clinton campaign. In a typical month, July, the top ten employers whose workers contributed to

Hillary Clinton included Google (number 1 by far), Microsoft (number 4), Facebook (number 5), Apple (number 6), and Amazon (number 10).

During the 2018 election campaigns, the employees of Amazon, Apple, Facebook, Google, and Microsoft made more than 125,000 donations to federal-level candidates for a total of more than $15 million. Only 1 percent went to Republicans, according to an analysis by *Wired* magazine. President Trump wasn't running, but his campaign did bring in all of $2,400.

This lopsided liberal slant shows up in the decisions these companies make in running social platforms for their millions of customers. It turns out that Google, Facebook, and Twitter all routinely censor things, and always to the benefit of the Left. They potentially have the power to influence American voters by using myriad manipulations, algorithm tweaks, and psychological tricks.

They practice this black art utterly free of any laws or regulations regarding their tricks, and without any public or internal guidelines restraining their behavior. They do so without disclosing any of their activities or the extent of them and without describing the effects on their users.

We know this because Facebook has conducted experiments on thousands and even millions of Facebook followers without telling them about it. It fiddled with their news feeds to see if a stream of positive versus negative items would make them any happier or sadder than usual (answer: yes), and it ran a voter test on 60 million "friends."

The New Oligarchs pose a clear and present danger to our nation's independent and free elections. We cannot place our trust in these companies, these people. They have failed to earn it, and in some cases, they have utterly abused it.

The mainstream media might otherwise be in frantic pursuit of this story, but the social media platforms provide extra circulation for media outlets. Plus, the New Oligarchs and other global billionaires have made dramatic inroads into the ownership of some venerable American media companies.

Amazon CEO founder Jeff Bezos bought the *Washington Post* in October 2013 from its longtime family owners, the Grahams. It is one of only three national newspapers with clout today, the *New York Times* and the *Wall Street Journal* being the other two. (Apologies: *USA Today* no longer counts for much.) For it Bezos paid $250 million, less than 0.002 percent of his wealth. Barely even a rounding error.

Bezos has avoided Twitter-trashing President Trump. On December 7, 2015, he took a snippy shot at the candidate six months after Trump had announced his run for president:

Finally trashed by @realDonaldTrump. Will still reserve him a seat on the Blue Origin rocket. #sendDonaldtospace http://bit.ly/lOpyW5N

The web link takes you to a YouTube video of a Bezos rocket blasting off—and landing vertically back on Earth.

Bezos's ownership of the *Washington Post* ensures him and Amazon the support of the most important media outlet in Washington, DC, the most important center of power that can hurt or help business. Amazon accounts for 40 to 50 percent of all online retail sales in the United States, and its cloud computing business hosts websites and back-office storage for hundreds of large companies. Now it combines this awesome market power with media cover.

Since President Trump assumed office, the *Post* has been his relentless and embittered opponent on its news pages and op-ed pages alike. Far more so than when the paper opposed both Bushes and President

Reagan. The president has returned fire on Twitter and elsewhere, targeting a threefer: the *Post*, Jeff Bezos, and Amazon.

The paper won a Pulitzer Prize for its reporting on Russiagate: on possible collusion by the president and his advisors, rather than an exposé of how the Deep State had staged a coup attempt to oust a newly elected president. Shameful. The Pulitzer Prize Board should rescind the prize, and the same goes for the prize that went to the *New York Times*, in the very same vein and for the very same reporting. The *Times* is 17 percent owned by the Mexican billionaire Carlos Slim Helú.

Meanwhile, *Time* magazine, one of only two national newsweeklies, was acquired in September 2018 for $190 million by the founder-CEO of Salesforce, Marc Benioff (net worth: $8 billion). Benioff, to his credit, withdrew himself from public comment on political issues to avoid undermining the magazine's claim to editorial independence.

Time's chief rival, *Newsweek*, was acquired in 2010 from the *Post* by the billionaire Sidney Harmon. He merged it with the Daily Beast, owned by IAC and its founder, the liberal billionaire Barry Diller. Harmon died a year later. Diller later said he regretted ever having bought the magazine, that it had been "a fool's errand." He sold it in 2013 to a small US media outfit, IBT. Its struggles continue; it clings to a bare existence.

The nation's two major business magazines, it should be noted, are owned by Asian interests who are likely beneficiaries of the free-trade and globalist policies that have hurt the United States so much for so many years. *Forbes* is owned by Wayne Hsieh, the founder of ASUSTeK Computer in Taiwan, and a partner in Hong Kong. *Fortune* was acquired in November 2018 by the Thai billionaire Chatchaval Jiaravanon, becoming part of a family empire worth upward of $40 billion.

• • •

In the Silicon Valley of the post-Obama era, a fiendishly liberal ethos combines with an assiduous eagerness to apply it to the business of search, social media, advertising, and entertainment. The current-day Valley is full of left-wing lunatics who have free rein to run wild while their employers indulge them.

At Google, the men's restrooms feature free tampons because, according to the HR department, "some men menstruate." The few conservative employees at these companies keep their heads down and fear being blacklisted. Multiple Googlers have said they faced discrimination for being white or Asian men, who are "overrepresented" in its workforce.

In the aftermath of President Trump's stunning upset in the 2016 election, senior executives at Google's headquarters campus in Mountain View, California, hosted an all-hands meeting with employees to soothe their devastated feelings. A tape of the mass session, part funeral and part group therapy, was leaked, and Breitbart published details.

"The Google tape" featured execs speaking to the troops, some of them choking back tears about how terrible it was that Hillary had lost an election they had considered to be such a sure thing. One influential Google vice president, Kent Walker, told the inconsolable crowd:

> *History teaches us that there are periods of populism, of nationalism that rise up, and that's all the more reason we need to be in the arena. That's why we have to work so hard to ensure that it doesn't turn into a World War or something catastrophic, but instead is a blip, is a hiccup.*

This is the comforting message from the veep of an online search company that suddenly sounds like a fifth column for the Bernie

Sanders campaign. The leaders of Google, when speaking privately to their staff, plainly state that their mission is to turn the Trump presidency into a "hiccup" on the long march toward "progress."

Walker again: "And yet, we do think that history is on our side in a profound and important way . . . the moral arc of history is long but it bends toward progress."

The leaked recording captured the lefty activist slant of Silicon Valley millennials and how managers must placate them. The Silicon Valley of old was based on merit: if you were the best in the world at what you did, it didn't matter who you were or what your color or creed might be. That idea is completely out the window.

Six days after the election, Salesforce CEO Benioff did an onstage interview at a tech conference in San Francisco. Speaking to Kara Swisher of Recode, a well-known and ardently liberal opinion slinger in Valley circles, he tried to comfort colleagues and put an optimistic spin on the Trump win. As Geekwire reported hours later:

> *Despite his wishful predictions that Donald Trump would never become president, Salesforce CEO Marc Benioff suggested on Monday evening that people forget their anxiety and give Trump a chance—at least initially. . . .*
>
> *"Over the weekend, I've seen a wide variety of emotions from my friends—from tears to anger to anxiety and stress—and we have to have a reset, let go of our fears. . . . There's a rhetoric when you're campaigning, and now there's an adjustment. He [Trump] said President Obama is actually a great guy. So there's a lot of positive stuff coming. I'm going to keep that position for now."*

If only they had listened. Kara Swisher would become a constant, caustic critic of the president, pelting him on Twitter repeatedly. Samples from December 2019:

Not that it matters to his dopey base, but: More than 500 law
professors say Trump committed "impeachable conduct."

And this one, with a flourish:

Mix rancid old tricks from running a sketchy real estate biz
mixed with a bilious topping of toxic Twitter and a bloated pile of
inaccuracies enough to sink a warlock to the bottom of the sea.

At Google, Facebook, and Twitter, the potential for abuse is heightened by the companies' unprecedented scale and reach. It is multiplied by the almost addictive connection they encourage from their
followers.

The internet sprang to life based on decentralization; the idea was
to create a computer network that could withstand a nuclear attack.
Silicon Valley has tossed that idea aside in favor of centralization and
the fearsome duopoly of Google and Facebook.

Facebook has subsumed many of the upstarts that might have
threatened its dominance. Google runs the biggest sites on the planet.
The two companies have a stranglehold on the online ad business.
Amazon, meanwhile, leads the world in cloud computing, serving
thousands of businesses and their websites. These days, if Amazon's
or Google's cloud services were to go down, a considerable portion of
the net would go dark.

Donald Trump is a prolific user of Twitter, and his techniques will
be studied by historians and social media researchers for decades to
come. He uses the platform to drive discussion, preempt and end-run
the Fake News, and shape the news cycle. Yet the hyperlefties at Twitter arguably hate everything he and the MAGA revolution stand for.
If he weren't the most influential and closely watched person on the
platform, they would have closed his account by now.

The platform's intense dislike of the tweeter in chief plays out in multiple ways. When the president retweets a post from a random person to relay it to his millions of followers, that person is often banned by Twitter. It is akin to pro sports: when a team's owners are angry at the star player, they can't fire him, so they fire his coach.

Twitter commonly "shadow bans" leading conservative voices, suppressing who sees their tweets, while also openly shutting down prominent accounts for hours or weeks at a time for purported violations of Twitter rules. Liberals run wild on the same site.

Twitter keeps creating rules designed to fight what it deems to be hate speech and misinformation. The problem is that the people setting the standards are purple-haired progressives in San Francisco who believe that Donald Trump is a Nazi. Those content moderators were fine with threats of violence against the Covington kids, but they will ban you for hate speech if you tell a laid-off journalist to "learn to code." That dismissive advice was the mainstream media's message to coal miners and factory workers in the Obama years.

When Twitter was just getting started in 2012, one of its executives, Tony Wang, general manager of the UK, described Twitter as "the free-speech wing of the free-speech party." CEO Jack Dorsey recently claimed that it was a joke. Which was quite true.

In 2018, Dorsey said that the social media site had unfairly reduced the visibility of 600,000 accounts, including those of some members of Congress. He told the BBC, "Twitter does not use political ideology to make any decisions, whether related to ranking content on our service or how we enforce our rules." He blamed problems on mistakes in algorithm design. Those mistakes always seem to hurt conservatives.

In 2019, an analysis of President Trump's Twitter account showed that the "likes" and retweets of the president's messages had fallen significantly. That was even more surprising, given the 32 percent rise to 62 million people following his personal account, @realDonaldTrump, since his inauguration. Interactions should have risen, according to the website Quartz, which did the analysis.

The president was sufficiently perturbed by that to summon Jack Dorsey to the White House for a meeting in April 2019 to talk about how the platform could help the nation respond to the opioid epidemic. The president also wanted to express his concerns over how Twitter was treating him and his followers.

On the day their meeting was to occur, April 23, 2019, the president began negotiating on Twitter ahead of time. He issued the first tweet at 6:26 a.m.:

> "The best thing ever to happen to Twitter is Donald Trump."
> @MariaBartiromo So true, but they don't treat me well as a
> Republican. Very discriminatory, hard for people to sign on.
> Constantly taking people off list. Big complaints from many people.
> Different names-over 100 M. . . . .

With a continuation at 6:32 a.m.:

> . . . . . But should be much higher than that if Twitter wasn't playing
> their political games. No wonder Congress wants to get involved—
> and they should. Must be more, and fairer, companies to get out the
> WORD!

The same month, the House Judiciary Committee heard testimony from Diamond and Silk, two flamboyant, voluble, Trump-supporting video bloggers who said their Twitter following had been suppressed by the site. That led to Senate scrutiny. Senator Ted Cruz held a hearing with testimony from Facebook and Twitter executives on whether censorship of conservatives was ongoing.

Earlier in April, Senator Cruz had told *The Hill*, "Big Tech behaves like the only acceptable views are those on the far left. And any views to the contrary are suitable for censorship and silencing." Of course, the oligarchs denied that.

A month later, the White House struck back. In May 2019, the administration launched an online tool for social media followers to use if they felt they had been wrongly censored, suspended, or otherwise banned from Facebook, Twitter, or another platform. The tool asks users for screenshots and links to the offending items and for details on which platform it occurred and any enforcement actions that were taken.

Twitter has fewer users than the other big platforms, but they include a higher proportion of media, political and business figures, and social justice warrior activists. Facebook's following is more than twice as large as Twitter's in the United States and almost six times as large worldwide.

Facebook uses "deboosting" to stop unwanted videos from going viral. As Project Veritas reported, an algorithm designed to block videos of violent crimes was also used to block some conservative videos and personalities, such as the humorist Steven Crowder.

In early 2018, Facebook engineered an algorithm change aimed at emphasizing mainstream media companies' content and pushing down the fare from narrower websites. That caused a 45 percent plunge in visits to President Trump's Facebook page and a 27 percent drop in visits to conservative pages. Meanwhile, left-wing sites saw no change in visitors or an increase of up to 12 percent.

Before the US midterm elections in 2018, Facebook banned 800 "antiestablishment" pages; it also blocked 30,000 people's accounts before the French elections and engaged in similar behavior in Brazil before the election of Jair Bolsonaro to the presidency. No outside overseer is examining the details and patterns regarding which accounts Facebook bans and which ones it favors. Facebook acts alone and in the dark.

This new role as censor arose out of Facebook's response to a hailstorm of bitter recriminations blaming it for Hillary's loss after a few pro-Trump ads appeared on the site. Russian agents had spent less than a million dollars on the ads. If that swung the election, it would make Facebook the greatest ad spend of all time. It was just more silliness from the manipulative and cynical Left.

The troubling thing is that Facebook's new role as self-appointed arbiter of what is appropriate to run on the site runs contrary to federal law (see chapter 14).

In early 2020, Facebook removed ads from the Trump campaign that it deemed to be misleading. The ads urged people to fill out the 2020 census form: "President Trump needs you to take the Official 2020 Congressional District Census today. We need to hear from you before the most important election in American history." They were urged to "respond NOW" and text "TRUMP to 8022."

Facebook cited a picayune problem: the ads said "today" when the census survey would begin in a week. In fact, that would hurt no one. The *New York Times* approved, reporting in a story on March 5, 2020, that Facebook was taking "a stand against disinformation ahead of the decennial population count that begins next week."

In the same month, Twitter deleted a tweet from Trump advisor Rudy Giuliani. Twitter decreed that the message had overstated the efficacy of hydroxychloroquine, the inexpensive malaria drug that the president had hailed. A similar tweet from my colleague at Fox News Laura Ingraham was also removed. It was a strange ruling, given the freedom that Twitter typically grants to liberals and our rivals and enemies overseas.

Therein lies the fatal flaw of the unbridled liberalism of these platforms: How can we be sure the petty, lefty censors at Twitter removed the Giuliani and Ingraham tweets for sound reasons, rather than because it was a way to oppose a president they hated? Once we know about the platforms' liberal bent, it casts suspicion on every regular business decision they make. They are tainted.

Google was founded twenty-five years ago, based on the mission of rounding up and organizing all of the information in the world. It was, in the main, an advertising company, and ad sales online provided 95 percent of its business.

That remains true today, yet Google (and its parent holding company, Alphabet) has grown up into far more than that. It is moving more toward curating the world's information—and serving it up in a style that suits its own purposes and its own political outlook. That style is knee-jerk liberal, sometimes to a ridiculous degree.

In 2019, Google formed an ethics advisory panel on artificial intelligence (AI), stocking it with engineers, policy advisors, philosophers, and other experts. As a token nod to conservatives, the company also named the president of the conservative Heritage Foundation, Kay Coles James, to the council. That prompted more than 2,000 Google employees to sign a petition to eliminate her from the AI panel.

The company's internal online chat boards filled with a freak-out fest. Breitbart and the Daily Caller published some of the comments:

Would we even consider having a virulent anti-semite on the advisory board? How about an avowed racist or white supremacist? This seems like a double standard where anti-LGTBQ positions are tolerated more than other extreme discriminatory views.

You don't need racists, white supremacists, exterminationists on the board to know their stances. You can just talk to their targets.

It's so upsetting that some of our leaders overlooked such hateful positions as Kay Cole [sic] James and the Heritage foundation have articulated and regularly advocate for.

Another employee suggested that the "rhetorical violence" of the Heritage Foundation "translate[s] to real, material violence against trans people, particularly trans women of color."

That last one is interesting. Does Google really care about "women of color"? Kay Coles James is a black woman. Though she had the right to speak and think—especially when invited by Google to visit its campus and do just that—Google employees sought to silence her because her views were more conservative than theirs.

If that had been done to a liberal black woman speaker by, say, factory workers in the Midwest, the left-wing media would have turned it into an indictment of President Trump and the racist wave he was fueling across the country.

Instead of Google senior executives' telling the staff to buck up and accept the right of divergent views to be heard, they caved. The invitation was rescinded, and Google scuttled the AI panel altogether. Here is what Kay Coles James herself had to say about the fiasco in an op-ed she published on April 9, 2019, in the *Washington Post*:

Unfortunately, some individuals inside and outside the company didn't share this appreciation for a diversity of viewpoints. That became clear last month after the members of the advisory council were announced. Some Google employees were so alarmed by the prospect of a conservative invading their playground that they started a petition to have me removed from the panel. It gained more than 2,500 signatures.

But the Google employees didn't just attempt to remove me; they greeted the news of my appointment to the council with name-calling and character assassination. They called me anti-immigrant and anti-LGBTQ and a bigot. That was an odd one, because I'm a 69-year-old black woman who grew up fighting segregation.

Last week, less than two weeks after the AI advisory council was announced, Google disbanded it. The company has given in to the mentality of a rage mob. How can Google now expect conservatives to defend it against anti-business policies from the left that might threaten its very existence?

Google's employees are so crazily far left that all Americans should shudder at the possibility that they are censoring the information that reaches us and tailoring it to their extremist kink. The company's efforts in censoring content—and the arrogance and imperious attitude of some of the troops—were revealed when Breitbart published a story on an internal report at Google in October 2018. Running eighty-five pages with the requisite glut of slides, it was titled "The Good Censor." Meaning: Google.

The report shimmers with the language of the Left and recent college grads. It opens by saying that Google, Facebook, and Twitter now "control the majority of our online conversations." Note that word: control. Rather than host or enable or facilitate—they control our conversations. It reveals more than was intended.

Also, the internal document admitted something that Google, Facebook, and Twitter have publicly avoided saying: that they have undergone "a shift toward censorship" in response to recent events. Read: the election of Donald Trump.

"The Good Censor" described Google's intentional shift away from the "American tradition," which the report said, "prioritizes free speech for democracy, not civility." The new approach was more like the "European tradition," which isn't free speech at all. The European model "favors dignity over liberty and civility over freedom."

Americans favor the First Amendment and unfettered free speech, even if it hurts the sensibilities of the aggrieved, the weak, and the oversensitive. Toughen up, guys.

Google and its sibling YouTube (the number one and number two most visited websites on the net) and Facebook and Twitter are torn between the "unmediated marketplace of ideas" and the need for "well-ordered spaces for safety and civility," the report states.

Safe "spaces"? The research paper cited bad fallout as a result of a no-holds-barred debate online: the rise of an ultraconservative political party in Germany—and the election of Donald Trump. In two of the slides of "The Good Censor," factoids about Russian meddling in US elections are juxtaposed on-screen with photographs of President Trump. According to Breitbart, "At one point, the document admits that tech platforms are changing their policies to pre-empt congressional action on foreign interference."

The idea of Google and other socialist social media heavyweights acting as self-censors is especially surreal in the context of an election campaign. Their interference would be easy to wage and easy to hide, and they could have a profound impact by fiddling a little with search results.

In the 2016 campaign, controversy raged over Hillary Clinton's 30,000 emails on a private server. Candidate Trump turned "Crooked Hillary" into a viral meme. Someone at Google rigged the search engine's algorithms so that when users started to type in her name, it filled in the phrase with "Hillary Clinton health care plan" or another harmless entry, rather than, say, "Hillary Clinton Crooked Hillary."

That was pernicious and insidious. It happens all the time.

In July 2019, a Senate subcommittee began looking into these issues and heard stunning testimony from a research psychologist who had spent almost seven years studying the practices of Google, Facebook, and Twitter. The Senate witness was Dr. Robert Epstein, a senior research psychologist at the American Institute for Behavioral Research and Technology. He was formerly the editor in chief of *Psychology Today*.

He told the senators at the hearing that he had been "a very strong, public supporter" of Hillary Clinton in the 2016 campaign. Yet he found it disturbing that the online platforms had been so biased in her favor. They had operated without disclosure or oversight, without the public's having any idea they were doing it and without anyone's knowing to what extent they were interfering in the elections.

Facebook and the other platforms were well aware of their inherent power because they had studied it in conducting experiments on millions of their own users—without telling them about it.

In Dr. Epstein's testimony on this to a Senate judiciary subcommittee on July 17, 2019, Republican senator Ted Cruz of Texas set him up:

CRUZ: Your testimony is that Google is, through bias in search results, manipulating voters in a way they're not aware of?

EPSTEIN: On a massive scale, and what I'm saying is, I believe in democracy, and I believe in free and fair elections more than I have any kind of allegiance to a candidate or a party.

CRUZ: And looking forward, if I understand your testimony correctly, you said in subsequent elections Google and Facebook and Twitter and Big Tech's manipulation could manipulate as many as 15 million votes in a subsequent election?

EPSTEIN: In 2020, if all these companies are supporting the same candidate, there are 15 million votes on the line that can be shifted without people's knowledge and without leaving a paper trail for authorities to trace.

The psychologist added a startling finding: if Facebook had posted a "Go vote" reminder in the news streams of only Democrats on election day 2016, it could have provided almost half a million extra votes by people who would otherwise have stayed home. "And we know this without doubt, because of Facebook's own published data."

In 2010, Dr. Epstein testified, Facebook had run an experiment "that they didn't tell anyone about," sending a "Go vote" reminder to 60 million users on election day and tracked the results. In all, 360,000 extra voters cast ballots that day, people who otherwise would have stayed on the sofa. Dr. Epstein told the subcommittee:

I don't think that Mr. Zuckerberg sent out that reminder in 2016. I think he was overconfident. I think Google was overconfident. All these companies were. I don't think he sent that out. Without monitoring systems in place, we'll never know what these companies are doing.

But the point is, in 2018, I'm sure they were more aggressive, we have lots of data to support that. And in 2020, you can bet that all of these companies are going to go all out. And the methods that they're using are invisible. They are subliminal, they are more powerful than most any effects I've seen in the behavioral sciences, and I've been in it for almost 40 years.

The delicate daffodils of Silicon Valley are so far left that they oppose even their own employers' strategic efforts if it means doing business with the demon Donald Trump.

In early 2020, Larry Ellison, the Valley legend who founded the database software giant Oracle Corporation and now is one of the five or ten richest people in the world, hosted a $250,000-a-couple fundraiser for President Trump at his estate at Porcupine Creek in Rancho

Mirage, California. That sparked a hail of protests from Oracle workers, who posted a petition on Change.org urging Ellison to cancel the event.

It said, "We are disappointed that Oracle Founder and CTO Larry Ellison's support of Donald Trump does not affirm Oracle's core values of diversity, inclusiveness, and ethical business conduct." Whatever happened to the core values of serving customers, maximizing shareholder returns, and making great products that improve business and the world? And America first.

Just shy of 10,000 people—or, at least, 10,000 accounts—signed the online petition. The show went on anyway. For Ellison, it was smart business. He had founded Oracle in 1977 and lately had been an avowed enemy of Google. He hadn't contributed to the Trump campaign in 2016, but now he was, as a *Fortune* headline put it, making "friends with his enemy's enemy."

For Ellison, it was a double bonus: Oracle also was a foe of Amazon, another Trump target. They are rivals in cloud computing. In 2018, Oracle had quietly formed a supposed grassroots group to attack Amazon. It issued 154 Amazon-bashing press releases in a year's time.

The New Oligarchs of Silicon Valley are so powerful, their resources so vast, that competition and the free market offer little promise of balancing their bias and their leftward lean. The incumbent platforms are so far out in front of competitors that they are going to be extremely difficult to catch. If it can be done at all.

This will make the antitrust angle all the more critical in seeking to restrain big tech from its most damaging instincts. The question is whether it is already too late to be able to do much to rein in Silicon Valley. Undaunted, President Trump, the Republicans, the Justice Department, and most of the nation's state attorneys general are already investigating big tech and searching for the right fixes and reforms.

@Jack and Twitter and the other oligarchs of social media had started this fight and picked a side, rather than staying out of it as they should have done, by law and in terms of a smarter strategy. Donald Trump was having none of it, and he decided to take the fight to them. Soon they would begin to feel it, bigly.

CHAPTER FOURTEEN

BREAK THEM UP

C onservatives love extolling the wonders of capitalism and free markets as a panacea for almost any malady. Their faith is warranted in a lot of cases—but a lot more than that will be required to rein in the New Oligarchs of Silicon Valley.

Something must be done to restrain their liberal excesses and expose their manipulative meddling in the information flow and political discourse of millions of Americans. Google, Facebook, and Twitter wield three formidable assets that make them all but impervious to any real competitive challenge: a dominant global platform, a clear and abiding lib bias plied without reservation, and the awesome power and the willingness to crush new entrants onto their turf.

At the same time, these are the New Oligarchs by default. No one checked their rise. We, the American people, and our leaders have only ourselves to blame for letting these hopelessly woke companies dominate our online lives and take over the internet.

The Congress of the United States is tremulous when confronted by a single one of these supertech entities. Most members of Congress failed to grasp their business models and how they managed to forge an oligopoly across businesses beyond tech and into news media, video entertainment, book publishing, online retail, job search, email, and more.

The immense business success of tech in Silicon Valley and beyond owes only in part to innovation and a preternatural sense of Darwinian competition. Big tech also relies on a quivering regulatory regime. This particularly is true at the FAANGs (Facebook, Amazon, Apple, Netflix, and Google). And Microsoft, a survivor of a government antitrust breakup bid in the late 1990s. Federal agencies stop short of trying, in any way, to assert national interests against the powerful interests of big tech.

Why the reluctance? Because, in many ways, the online platforms have become vital to the very people who otherwise might want to rein in their predatory and anticompetitive modus operandi: congressmen, senators, regulators, and the mainstream media across the entire political spectrum.

The old advice to politicians on taking on newspapers was to never pick a fight with a man who buys ink by the barrel and paper by the ton. Today few politicians want to pick a fight with the giants that possess the pixels. President Trump, intrinsically and characteristically, was willing to do that.

The reluctance to take on big tech also is shared by the mainstream media. They were largely silent on the looming influence of the big tech trio, in part because they all shared an ardent liberal outlook and an antipathy for President Trump.

The Fourth Estate is, without question, in ruins. The national left-wing media have cratered under the weight of their political bias and the loss of billions of dollars in ad revenue to their online successors. Local news is near extinction. Social media, meanwhile, are ascendant—and the frail national media are more reliant than ever on Google, Facebook, and Twitter for their circulation and profile.

Can the *New York Times* really risk launching a deep investigative report on the threat to democracy posed by the bias of the three social media giants?

Yet if big tech is so terrible, the opportunity should be wide open for someone to build something better. Like the argument for free trade with foreign countries, this view breaks down in practice.

In the old Silicon Valley, the heavyweights welcomed the fledglings and cheered their rise. In today's meaner clime, the Valley is hostile to innovation and start-ups. The giants buy or crush newcomers. They hold enormous market share in multiple areas with the ability to bar new competitors—and they do so whenever it suits them.

In the eight years of the Obama administration, Google grew more than fourfold, from $21.8 billion in annual sales in 2008 to $89.98 billion in 2016. It shifted from being just a search engine company to being a dominant email provider (Gmail has more than 1.5 billion users), the owner of the number one and number two websites in the world (Google and YouTube, respectively), and a provider to thousands of schools filled with future customers.

Your children or grandchildren might be completing their lockdown schoolwork using Google Classroom. They might do their homework via Google Docs, a "free" word processing program that the internet giant actively monitors.

Every day, Google runs 3.5 billion online searches for users around the world. Its search rankings can make or break a business. It controls 85 percent of the worldwide search market. Its Android software runs a massive 75 percent of the worldwide market for smartphones. Apple's iOS runs 24 percent, so together the two oligarchs own 99 percent of the global market for the most dynamic, most personal technology product ever.

If Apple and Google deny your app admission to their app stores, they just blocked you from connecting with 99 percent of the smartphone customers in the world.

Facebook has 1.6 billion daily mobile users and 2.3 billion monthly mobile users globally. In the United States, it holds collectively almost half a billion overlapping accounts: 190 million Facebook "friends," 107 million users of Instagram (of 1 billion worldwide), 68 million users on WhatsApp (of 1.5 billion worldwide).

Much of this expansion was funded by Facebook funny money in the form of a soaring stock price. It acquired seventy companies in fifteen years. The acquisitions of the nascent Instagram photo-sharing app in April 2012 and the WhatsApp messaging app in February 2014 let Facebook gobble up its two biggest competitive threats at the time.

The deal for Instagram looked crazy and turned out to be one of the biggest high-tech payoffs of all time. Mark Zuckerberg, Facebook's founder and CEO, had spied the photo-sharing app when it was a thirteen-person start-up with zero revenue and 30 million customers. He paid $1 billion in stock and cash for Instagram in April 2012, a month before Facebook went public.

In the next eight years, Facebook's stock price rose fivefold to $215 a share, and its market cap grew past $600 billion. Instagram's user base rose *thirty-three-fold* in the same period. Similarly, Facebook paid $16 billion for WhatsApp in 2014, and the number of its users has more than tripled to 1.5 billion in the years since.

The duopoly of Google and Facebook in online and mobile advertising is staggering in wealth and scale. Google controls almost 80 percent of worldwide ad spending for online search. Facebook holds a similar share of the worldwide market for ads on social media. The two collect more ad revenue, combined, than all spending on all television advertising worldwide: $230 billion projected for 2020, compared with $190 billion for television.

In the United States, the pair controls 60 percent of online ad sales and Amazon has 10 percent, leaving only 30 percent of the market to be divvied up among hundreds or thousands of other advertising companies.

This gives the two titans a grip on the future of advertising, which funds the future of content and media. Traditional advertising was forecast to grow by only 1.5 percent to approach $325 billion in 2020, while the online market was set to grow much faster to reach an even bigger number: up by 13 percent to $335.4 billion, according to the market researcher WARC.

Twitter has 330 million active users worldwide and more than 80 million in the United States. Netflix has 180 million customers worldwide. Amazon has 112 million subscribers to its Prime Video service. Apple has almost three-quarters of a billion iPhone customers and an app store where they can sample from among 2.2 million apps.

Google and Facebook, in particular, are fiendishly vigilant in thwarting, crippling, or acquiring any possible competitive threats hatched among the venture capital seedlings planted in Silicon Valley. For all their dominant market power and the vast lead they have over rivals, they live in paranoid fear that someone is gaining on them, an endemic Valley trait that predates the liberal contagion that has consumed the place.

Their predecessors atop high-tech made software and hardware that wears out in ever-faster product cycles. This provides openings for innovation and faster-better-cheaper newcomers. In social media, the opportunities are fewer. There have been no real product upgrade cycles and technological leaps in the way people talk to one another online. Once more than 2 billion people around the world are on Facebook, the social site has reached escape velocity—there is almost no way for any other service to supplant it.

Thus, venture capital investors want to place their tech bets elsewhere. It leads entrepreneurs and product creators to focus else-where, often on making new tools and apps that leverage off the Face-book and Google platforms. The giants have become ecosystems, making the rest of tech more dependent on their beneficence rather than intent on toppling them.

Google, Facebook, and Twitter really don't offer a product that rivals can outmatch. They are entire worlds unto themselves, all-encompassing online environments where all the commerce, life, and exchanges that occur can do so only because these platform overlords have enabled it—and because they allow them to continue.

Google is a paragon of the monopolistic market bully that govern-ment is usually eager to bust up. It has been accused of crushing com-petitors or stealing from them when it is unable to out-innovate them or acquire them. Its self-identifying posture as a left-leaning beacon of Silicon Valley is, in part, a matter of smart business.

Its workforce of young millennials demands it. Its embrace of the Obama administration spared Google any serious antitrust scrutiny in the United States for eight years. Obama owed much to Google for his electoral success. His administration let Google do whatever it wanted to do. Meanwhile, regulators in Europe slapped the company with record-setting fines exceeding $8 billion for abuses in wringing every last drop of data out of its users.

In 2012, the staff of the Federal Trade Commission recom-mended filing antitrust charges against Google for abusive tactics that "resulted—and will result—in real harm to consumers and to innovation in the online search and advertising markets." The staff recommendation resulted from an investigation over the previous twenty-one months.

Google lobbyists increased their visits to the White House in that period, as the *Journal* later reported. A year later, President Obama's appointees to the FTC voted unanimously to overrule the staff recommendation and forgo filing any charges.

Despite its massive scale, Google continues to scrap like a barroom brawler. Restaurant review site Yelp has accused Google of scraping Yelp listings and absorbing them into Google's search results in an attempt to divert consumers from visiting Yelp directly. In targeting travel websites, Google pulled the same tricks, and later it moved into online job searches.

In other cases, it has been accused of stealing technology. The smart-speaker company Sonos, saying that Google had pilfered its technology, filed a patent infringement lawsuit over it. Oracle has won a lawsuit claiming that Google stole its Java language and used it in the operating software for Google's Android phones made by companies around the world. Google's appeal of that case was bound for the Supreme Court in 2020.

A lot of competitors are too small to fight back effectively, but never too small to get shut out of the market by the big guys. This is shown by the story of Gab, a new social media site that billed itself as a free-speech alternative to Twitter. It offered features Twitter lacked, such as the ability to edit posts after hitting "Send." Gab began picking up traction and then got crushed by the New Oligarchs. Google said it bans any apps that could be used by hate groups. Someone must have forgotten to tell it that Twitter hosts accounts for the American Nazi Party and the Ku Klux Klan.

Here were some of the hurdles thrown up in front of Gab: Google and Apple refused to add its app to their app stores. Credit card processing companies shut down Gab's ability to accept payments via card for memberships, donations, and merchandise. Its accounts with email services were terminated, stopping it from sending emails to members. Its domain name account was terminated, so that typing

"Gab.com" into a web browser would return an error instead of directing users to the site.

Every possible roadblock was placed in front of the tiny start-up. When the free-market crowd says, "Just make your own site," it ignores the fact that Apple and Google can block any new rival from access to the entire smartphone world. Gab survived those challenges but had to devote resources to reinventing its underlying software code instead of growing and marketing its business. Many other rivals have gone out of business.

Such anticompetitive behaviors should be a prime target for our antitrust laws. In 1984, the federal government forced the breakup of the old AT&T telephone monopoly. Phone prices were exorbitantly high, innovation was stymied, and competition was blunted by AT&T's heavy-handed tactics. Windfall benefits redounded for the next thirty years. On the very same day in 1984, it stopped short of breaking up IBM. Later, IBM was humbled and stripped of any predatory power over customers by frenetic competition and innovation rather than government intervention.

In the 1990s, in the case of *United States of America v. Microsoft Corporation*, which had been fomented by jealous rivals, the government won a key round that, now that we read about it, reveals the entire case for how spurious it was. A federal district judge ordered Microsoft to stop bundling its Internet Explorer web browser with its Windows operating system for PCs, so outfits such as Netscape would have a fair chance to compete. The government lost on appeal, the two parties settled, and Microsoft would go on to lose its unbridled monopolistic power to Google and the reborn Apple.

In the case of the New Oligarchs in the 2020s, the picture is more disturbing for its implications, antitrust and societal. Their dominance may last for decades into the Trump Century. The ideologies, messages, and voices they promote most on their platforms will thrive; those they ban could fade.

In 2019, investigating big tech came into vogue in a uniquely bipartisan way. In July, Congress held antitrust hearings in Washington and summoned executives from Facebook, Google, Amazon, and Apple, all of which were under Justice Department review.

Meanwhile, the Federal Trade Commission fined Facebook $5 billion for mishandling users' personal data, the largest fine the agency had ever assessed. Google agreed to pay $170 million in federal fines related to YouTube and child privacy.

On Friday, September 6, the attorneys general of eight states, led by New York and including four Democrats and four Republicans, put out a statement announcing a joint antitrust investigation of Facebook. The following Monday, several state AGs held a press conference in Washington to say that forty-eight states were launching an antitrust probe of Google. The two states declining to join the effort were California, the home base of Google and Facebook, and Alabama.

The new investigations were going deeper than the main issue of whether Facebook and Google are too big and too brutal in the online advertising market. The AGs were probing the two companies' handling of consumer data and their role as the gatekeeper of a lot of internet activity and online competition.

Antitrust investigations proceed only slowly over months or years. A more urgent pushback was necessary to curb the overly liberal bias of the online platforms. Conservative voices were the ones being muted by Facebook, Google, and Twitter. Better-known right-leaning figures might be more outraged—but some of them were on the Silicon Valley payroll.

One example is the *National Review*. Its editor in chief, Rich Lowry, came out in defense of the New Oligarchs in early 2019, publishing an op-ed on the liberal *Politico* website beneath the plaintive headline "Don't Break Up Big Tech." It was coming under fire from both the Left and the Right. Senator Elizabeth Warren, outraged that Facebook had rejected her ads *criticizing Facebook* on Facebook, called for the breakup. It was a strange reaction: Would she expect CNN or Fox News to be required to run commercials criticizing them on their own air?

Nonetheless, when Senator Warren tweeted with the umbrage she brings to most issues, the superlib was joined by a surprise ally: arch-conservative senator Ted Cruz. That prompted the *National Review*'s Lowry to write on *Politico* on March 13, 2019, "Tech is caught in a right-left pincer, made all the more powerful by the populist spirit afoot in both parties." He continued, "Conservatives don't like these companies because they are owned and operated by sanctimonious Silicon Valley liberals subject to the worst sort of groupthink. Progressives don't like them because they are colossal profit-making enterprises." True in both cases.

Lowry even praised the oligarchs of Silicon Valley as "the most successful and iconic American companies." There was just one thing Lowry forgot to mention: that Google funds the National Review Institute, which owns the *National Review*. So the Google-funded organization says we must leave Google alone. Now it makes sense.

Google funds various conservative think tanks in the same way any corporation funds most any cause, as a way to buy influence, access, and good PR. In fact, audio of a Google internal meeting leaked to Breitbart News shows senior execs all but laying this out during a meeting with the staff. It was an attempt to quell sensitive, triggered Googlers horrified by its support of anything rightward.

Many employees were upset that Google was a sponsor at the Conservative Political Action Conference (CPAC), a huge annual gathering in the Washington, DC, area. In late February 2018, Google sponsored what the *New York Times* later reported to be "a lavish hospitality suite, courting conservatives with an outdoor fireplace, hors d'oeuvres and flowing cocktails. Bright young representatives from Facebook hosted a 'help desk,' handing out cookies frosted with emoji icons."

That sparked an outcry among the hothouse flowers at Google's base in Mountain View. Senior management hosted an all-hands meeting to soothe the troops, and it was recorded—and leaked a full year later. In the session with employees, Google's senior director, U.S. public policy, Adam Kovacevich, admitted that CPAC was a "sideshow circus" and an important opportunity to "steer" the conservative movement. Read his words carefully:

> *The majority of Googlers would want to steer conservatives and Republicans more toward a message of liberty and freedom and away from the more sort of nationalistic, incendiary comments, nativist comments and things like that. But it has been a very valuable place for us to reach a lot of the people and the big tent of conservatism.*

Google's proper role, we thought, was to serve up listings of online information based on simple queries we type into its iconic search box. The thought that senior executives at the company arrogantly view their role as "steering" us anywhere is alarming—and 82 percent of the people on the internet worldwide use Google.

At the CPAC gathering a year later, from February 27 to March 2, 2019, Google and the other companies were no longer quite as welcome. As the *New York Times* reported on March 4:

> *At last week's gathering here in a suburb of Washington, Silicon Valley's only obvious presence was on the lips of exercised right-wing*

critics who whipped up the crowd by denouncing the American tech industry as an authoritarian hegemony intent on censoring their cause.

"Facebook, Google, and Twitter are pushing a left-wing social agenda while marshaling their marketing power to shut conservative voices out of the marketplace," said Senator Josh Hawley, a Missouri Republican, during a featured session with the ominous title "Blocked: This Panel Has Been Removed for Conservative Content."

Three days after that story ran, Breitbart broke the news on the leaked tape of the Google protest meeting sparked by the CPAC sponsorship a year earlier. It proved beyond all doubt that the conservatives had made the right call.

Another outfit against breaking up big tech is the Heritage Foundation, whose spokesman Robert Bluey has gone on record as decrying the "heavy-handed government regulation" of the internet giants. Unmentioned is the fact that the Heritage Foundation takes Google money.

Senator Hawley has harshly criticized the Heritage Foundation, along with other supposedly libertarian groups, for being apologists for big tech. The revelation of Google's link to the Heritage Foundation sparked an outcry from the staff so fierce that the company scuttled plans for a new AI panel whose members were to include the group's president.

So far, competitors have been unable to rein in GooFaTwit. The same goes for politicians, regulators, and the mainstream media. Who will fight them? Most times in the United States, this kind of gap lures the trial lawyers, who can find a multitude of inventive ways to attack a target and exploit its vulnerabilities.

This tack is all but impossible in the case of the social media triopoly. Under federal law, internet service providers and web hosts are classified as the equivalent of telephone carriers. The old AT&T and the

"Baby Bell" local phone spin-offs weren't to blame for the things said in phone calls on their networks and therefore they couldn't be sued for libel as a newspaper, magazine, or TV news channel could.

Nor could online providers, such as Comcast in cable, the local telephone companies, and the old America Online in internet access, be held liable. The rise of Facebook, YouTube, and others came a decade later, and they assumed the same setup.

This extra level of protection, higher than even the First Amendment provides to the media and the American people, was part of the Communications Decency Act of 1996. Congress had passed it when the World Wide Web was just getting started, and unbridled expansion was prized. Let a thousand flowers bloom.

The original aim of the CDA was to curb minors' access to pornography, which now accounts for 30 percent of all traffic on the net. It protected internet providers from getting sued for blocking obscene materials from reaching their audiences. It also ensured online carriers' freedom from legal liability and lawsuits on everything beyond porn.

That now undeserved protection lies in Section 230 of the CDA: "No provider or user of an interactive computer service shall be treated as the publisher or speaker of any information provided by another information content provider." Internet platforms were free of any liability for the words and posts of others on their platforms.

At the time, Section 230 made sense. The *New York Times* exercises total control over the words it prints on the news and opinion pages. If someone libels me there and the paper relays it to a million readers, both the author and the newspaper deserve to be sued. I would stop short of also suing Facebook when the *Times* posts the story there: that, too, was the paper's call. Facebook has less control over what 2 billion people say on its platform.

Section 230 was written into the CDA with a noble purpose: to promote free speech on the internet and *protect political speech from censorship*. Today, the law has been twisted. Now it protects Google, Facebook, and Twitter from any liability or responsibility, even though they are expanding their roles as censors and curators of the correct political views. This is in direct violation of the intent of the law.

The new law was supposed to protect neutral platforms, but Facebook, Google, and the other behemoths of Silicon Valley openly admitted that they were, instead, publishers making editorial decisions. Why, then, did they get the protection afforded to neutral platforms? Section 230 is now used as a shield for the oligarchs as they remove speech they dislike (usually conservative speech) without fear of legal repercussions.

It is a recipe for impunity. Google, Facebook, and Twitter have total authority and power over their platforms—without also bearing legal liability for their own actions or for what ensues there and without having to pay the cost of maintaining and upgrading the internet infrastructure that delivers their products to more than 2 billion people. The New Oligarchs are free to do as they like, and they and their employees wallow in too much Trump hate to be trusted.

In June 2019, Senator Josh Hawley filed a bill in the Senate to strip the oligarchs of Section 230 protection if outside inspections unearthed evidence of bias and underhanded tactics. He had started pursuing big tech when he was state attorney general in Missouri. His new Ending Support for Internet Censorship Act would impose bias audits on internet giants and inspect their algorithms and content moderation policies. For companies such as Google, Facebook, and Twitter, which have been caught red-handed at this so many times, even proposing such a law may make them think twice about blacklisting the next conservative voice.

Six months later, Attorney General William Barr escalated the attack on the New Oligarchs and their abuses of Section 230. It came in an address he made on December 10, 2019, to the National Association of Attorneys General at a gathering in Washington, DC. At the time, forty-eight state AGs, plus those of Puerto Rico and the District of Columbia, were already jointly investigating Google for antitrust violations. Barr told his colleagues:

> Section 230 has been interpreted quite broadly by the courts. Today, many are concerned that Section 230 immunity has been extended far beyond what Congress originally intended. Ironically, Section 230 has enabled platforms to absolve themselves completely of responsibility for policing their platforms, while blocking or removing third-party speech—including political speech—selectively, and with impunity.

Barr went on to describe for his state counterparts the various ways antitrust law could be applied to the social media platforms and the status of the Justice Department's ongoing antitrust review of Google's many bullying behaviors.

Big tech is also a complicit and willing ally of our biggest enemy and rival, China. Their social justice warrior workers decry President Trump, yet the tech giants bow and scrape to Chinese leaders in an effort to appease them and do more business with the country. Silicon Valley is happy to work with the brutal Communist Chinese government, even as its employees protest working with the US government.

Though Facebook and Google remove tweets from presidential advisors and Fox anchors, they blithely allow, without question, Chinese government ministers to propagate propaganda to millions of Americans. China asserted on Twitter that US military troops brought the Wuhan virus to Wuhan, and Twitter let that ridiculous accusation stand unchallenged. Google was developing a censored search engine for China called Dragonfly until it was shut down in July 2019. It was even building technology into the search engine to help China track dissidents. Don't be evil.

If you have teenage children or grandchildren, they may be on the social media app TikTok, where youngsters upload short video clips of themselves dancing and singing. That app is owned by China, and only it knows what level of information about your loved ones and their devices is being collected. The Communist Chinese are the only rivals of Silicon Valley in lacking any compunction about gathering up every detail of your life from apps and devices, whether for profit or for some nefarious purpose.

China's influence on Americans' internet and entertainment habits is spreading. Fortnite is the most popular video game in America for teens and younger children, and it has a quarter-billion users around the world. It is owned by Tencent, a giant Chinese gaming company. Epic, the US company that hosts Fortnite, claims it shares nothing with the Chinese company.

Then again, the oligarchs of Silicon Valley have testified to Congress that their platforms demonstrate no bias against conservatives.

Working with China seems an intolerable hypocrisy for the oligarchs. Apple deigns to dictate whether Americans can view particular ultraconservative opinions on their iPhones, which were made in Chinese plants where some workers have attempted suicide by jumping out of a nearby window because their working conditions were so mindless and bleak. The factories have erected nets to stop them from succeeding.

The next big battle with China will break out in the competition for world domination of the next-generation wireless design known as 5G, for "fifth generation." Right now, the network speed for smartphones is 4G, but we are rapidly approaching 5G networks, and we are lagging behind China. A big part of the problem has been Silicon Valley's reliance on the Chinese company Huawei for hardware. China experts argue that Huawei is the de facto technology arm of the Chinese surveillance state.

According to the China critic Gordon G. Chang, the United States must block Huawei technology from infiltrating US wireless networks. The gear would let China scoop up the world's data, eavesdrop on any communications it wanted to monitor, and control billions of devices that will be linked to the Internet of Things.

In February 2020, retired air force brigadier general Robert Spalding, a senior fellow at the Hudson Institute, warned that the Chinese government uses US companies to push its point of view on the US government. He said that US companies such as Qualcomm "come to the Department of Defense, seeming like an ally, when in reality they're working as a proxy for the Chinese Communist Party, because that's how China and the Chinese Communist Party has incentivized the system."

Also in February 2020, conservative groups including FreedomWorks, R Street, and the Lincoln Network sent a letter to the Federal Communications Commission on the urgency of the 5G rollout:

> Enabling rapid, widespread deployment of 5G wireless in urban and rural areas is critical to maintain America's economic well-being, national defense, and global competitiveness. We do not underestimate the stakes; China continues to advance on 5G and could surpass the U.S. in artificial intelligence, augmented reality, quantum computing, and robotics. The C-band is where 5G innovation in these technologies will happen, and if the U.S. does not provision this spectrum as quickly as possible in 2020, it will lose the opportunity to compete with China—a potentially devastating but avoidable outcome.

The issue of 5G is of global importance and vital to US interests—
and we must make sure we can trust Silicon Valley to abide by the
Trump doctrine of America First. So far, big tech has failed on this
point.

There is no one solution to reverse the course of Silicon Valley and
return it to its free speech roots—especially if the Democrats won the
2020 presidential election and the government was overtaken by the
Valley's ideological doppelgangers. If President Trump won reelection, we would have a chance to reclaim free expression and meritocracy in the hallways of US high-tech companies.

Democrats embrace the censorship of Silicon Valley, and any
Democratic administration would witness a return to the Obama
years, with a revolving door of tech employees in DC, an internet
policy dictated by the progressive lunatics at Google and Facebook,
and a red carpet rolled out for Communist China's Huawei.

There are ways to push the New Oligarchs back to the center and
true neutrality. An attack on Silicon Valley on many fronts would undercut its business model of exploiting private data and flaunting legal
protections. When profits fall, even the most progressive oligarch will
begin listening to customers again.

In a second term, President Trump might make significant inroads
in advancing on this front. A model for how he might proceed may lie
in another great victory he racked up in his first term, this one in June
2018: the repeal of net neutrality.

It was seen as the crown jewel of Obama's internet policy, and it
had Google's fingerprints all over it. The Democratic majority on the
Federal Communications Commission passed a new regulation restricting Comcast and other big carriers from blocking content from
other giants, including Netflix and Google's subsidiary YouTube—
something they had never done, to begin with. That was an issue that
Google had helped create; no real problem existed.

The rule, approved by the FCC on February 26, 2010, forced Google's carriers—internet service providers (ISPs)—to treat all internet traffic as equal. It banned them from demanding special deals even of companies whose streaming video chokes their bandwidth, Netflix and YouTube in particular; the two services hog upward of 20 percent of all internet bandwidth in the United States. Silicon Valley loved net neutrality because it restricted carriers, which had the burden of maintaining their networks while protecting content suppliers like themselves.

The Obama administration sold net neutrality to the world with scare tactics, saying that ISPs could censor you at any time and silence your voice on the internet. Sound familiar? It is what Silicon Valley does to conservatives daily.

Donald Trump installed a champion at the FCC in the form of Chairman Ajit Pai. Pai immediately moved to repeal net neutrality. On November 28, 2017, he proposed repealing the regulation and released a 3,000-word statement to his fellow FCC members, urging them to vote with him on the issue. He pointed out that although the regulation had been aimed at controlling the behavior of ISPs, it was "edge providers" such as Google, Facebook, and Twitter that were censoring people and their view. He wrote:

[Edge] providers routinely block or discriminate against content they don't like.

The examples from the past year alone are legion. App stores barring the doors to apps from even cigar aficionados because they are perceived to promote tobacco use. Streaming services restricting videos from the likes of conservative commentator Dennis Prager on subjects he considers "important to understanding American values." Algorithms that decide what content you see (or don't), but aren't disclosed themselves. Online platforms secretly editing certain users' comments.

*And of course, American companies caving to repressive foreign gov-
ernments' demands to block certain speech—conduct that would be
repugnant to free expression if it occurred within our borders.*

*In this way, edge providers are a much bigger actual threat to an
open Internet than broadband providers, especially when it comes to
discrimination on the basis of viewpoint.*

It was a scathing summary, and every word of it was true. In his
conclusion, he anticipated the virulent overreaction his words would
trigger among the sensitive radical lefties in Silicon Valley: "So when
you get past the wild accusations, fearmongering, and hysteria, here's
the boring bottom line: the plan to restore Internet freedom would
return us to the light touch, market-based approach under which the
Internet thrived. And that's why I'm asking my colleagues to vote for
it on December 14."

Pai's blunt broadside sent everyone else into a tizzy in Silicon Valley, the leftist mainstream media, and among the Dems. It was the end
of the internet, they declared. Free speech would die, and the internet
would slow to a crawl. That last part was the universal refrain: internet speeds would slow dramatically as costs ballooned because the
ISPs would feel they could get away with it.

President Trump and his new Republican FCC chairman got their
way, and the FCC repealed net neutrality on June 11, 2018. The internet kept working fine—in fact, it got faster than ever. The repeal
let the ISPs know that they had the power to avoid being forced into
bad deals and that they could recover more of their cost of expanding
the infrastructure of the internet by billing bandwidth hogs more than
normal users.

The internet has sped up dramatically since the repeal of net neutrality. According to Ookla, which tests internet speeds, broadband
speeds in the United States improved by 36 percent in the first year

after the repeal. They have likely gone up since then. Without burdensome net neutrality rules, the free market is working, something that should thrill conservative think tanks, even if they remain in Google's pocket.

It's truly Trumpian and a promising step toward returning Silicon Valley to its rightful place as a center of innovation, job creation, and free speech for all. Only Donald Trump was capable of winning that fight, and only if Americans gave him that opportunity.

EPILOGUE

Trump has proven in the course of his contentious presidency that he is the man for this era, and he stands alone in that distinction. He is a uniquely American leader: bright, bold, a tough fighter, tested and true to his values, a Yankee original. His qualities of character are obvious in the way he has run his administration—no entourages, no bloated staffs; he is a man who says what he means and means what he says, whether wrangling with Fake News, giving a speech on television, or holding forth on Twitter.

Even by billionaire standards, Trump is unusual: he's a builder, a developer, a promoter, a reality TV star turned politician with his own brand of populism. Trump supporters love their standard-bearer because he is partly like them: he works hard, he talks straight, and he's an unabashed optimist.

President Trump can be brash, sometimes frequently so. He is now the recognized champion of rough-and-tumble American free enterprise and capitalism. He is unafraid to thank God and pray for His help when the going gets tough—when the tests become vast, complex, and daunting even for him. When talking to the American people, he forgoes sugarcoating his language or his meaning. He talks in the direct, straightforward language of the plainspoken American, using short Anglo-Saxon words that carry meaning and weight and that resist obfuscation and create clarity.

This befuddles snarky radical Dems, cynical Fake News peddlers, and all those who believe only in Never: Never Trump, Never America, and Never God, Never Guns. President Trump is a devout believer in the possible. He's a doer, a man of action, and he loves to taunt the sanctimonious stooges of the Left and their pedestrian thinking.

I have never seen a survey of what the Fake News media expect of him, but I'd wager they never dreamed for a moment that he would be the one to return God to the public square and arena. Maybe they thought he would make the middle class and all who aspire to it fashionable once again. But did they really think he would build the net worth of minority households and cut minority unemployment to the lowest levels ever? And start bringing home some of the manufacturing jobs that corporate America shipped off to cheap foreign labor markets?

His presidential achievements, his accomplishments, are now the stuff of history. And for good measure. He has also reduced the number of our troops in foreign countries and rebuilt the military, while building peace.

President Trump had presided over an economy with the richest markets in our history, our economy growing at rates former President Obama once called impossible. He had accomplished all this while reminding all Americans how exceptional we are, how blessed we are. Now he is focused on what he calls the Transition to Greatness.

He is leading the nation out of the darkest days of the plague that China and Xi Jinping unleashed on an unsuspecting world. The Wuhan virus was a deadly contagion that infected more than a million of our fellow citizens and killed more than a hundred thousand Americans and hundreds of thousands more around the world.

Dozens of our best and brightest researchers and physicians in bioscience and medicine are racing to create a vaccine. There is no question that they will be successful because that's what Americans do, we succeed, and it took this president to remind us of this fact. President Trump leads by example. He never surrenders his faith in all that we can do and must do for this nation to succeed.

Incredibly, President Trump earned his place in history in just a little more than three years in office in his first term. And now we are asking him to do it all over again, and to do more. He was elected despite a conspiracy by the Deep State and radical Dems doing their worst to block him from winning the election of 2016. Then President Trump overcame three years of subversion, resistance, and obstruction by the RINOs and radical members of both parties, a criminally orchestrated investigation led by fools and traitors who were so brazen, arrogant, and corrupt that they tried to impeach a president who had just been exonerated by the special counsel.

The radical Dems' impeachment farce collapsed quickly, and on February 5, 2020, the Senate acquitted him of the spurious charges concocted by the Party of Hate.

While the Dems tried and failed for a second time to carry out a coup against the Trump presidency, he had already moved on to the next crisis: he had implemented a travel ban against China to stop the virus from migrating to the United States. The deadly pandemic was the next great battle of his first term, fighting against what he called the invisible enemy: the novel coronavirus, which causes the disease covid-19.

Or, as I will always call it, the Wuhan virus.

As usual, he began his fight all but alone. Democrats, the left-wing media, and indeed many members of his administration and his party ridiculed his judgment and his decision to enforce the ban against the entry of people who had visited China in recent weeks. No matter, the ban saved American lives, perhaps more than a million.

It was one of President Trump's greatest decisions. In the midst of crisis and a national emergency, he acted in the interest of all Americans, and against so-called expert second-guessers and, of course, the disloyal opposition and left-wing media. He may never get the full credit he deserves, or the thanks. But, as you probably noticed, President Trump would achieve great things even if he worked for only a thank you.

He does what he believes to be right for the United States. He speaks directly to his followers on Twitter and social media, and he speaks his truth. His followers, supporters, and voters love him for his fresh language, his direct and honest voice. There has never been a president like him.

Even as he campaigned in 2015 and 2016, the left-wing media and Republican establishment were shaken by his candor, as he attacked President Obama for negotiating the dumbest deal in history with the Iranians, and then also had the temerity to take on the Bush dynasty. Trump mocked George W. Bush and dismissed his brother Jeb by labeling him Low Energy Jeb, a tag that Jeb could never escape.

Trump ran over the field of sixteen GOP candidates and dominated the news with fresh takes and outrageous statements. And when he was on television, ratings soared. Cable networks couldn't resist the Trump Train, and the Trump Train just kept building speed while the forgotten man and woman of America reveled in the Trump spectacle and Trump became the man to beat.

All of his rivals and almost every left-wing news outlet tried to destroy the candidate they had helped create, but their usually reliable politics of personal destruction had little effect. In fact, they may actually have helped fuel the Trump candidacy.

They trotted out their usual labels. The media called Trump a racist because he wanted to build a wall on the Mexican border to stop illegal immigration; they screamed, "Xenophobe!" because he wanted to end the United States' decades of consecutive trade deficits with the world, and then they tried out "impossibly naive" because he wanted to stop outsourcing well-paying American jobs to cheap foreign labor markets.

I was among those cheering the president in every debate, in every rally, in every speech, because I shared his view of the importance of balanced trade and strong economic growth, the necessity of securing our borders to end illegal immigration, and to restore the United States' preeminence in innovation and manufacturing. Bring jobs home instead of allowing the Chamber of Commerce and the Business Roundtable to strip the members of the American middle class of their jobs and ship them to China, India, or Romania.

I also knew how it feels to be attacked by the cynical, bitter Left and the global elites of the business establishment. President Trump offended the same Washington establishment and the elites of both parties, and I cheered louder than ever for him when it became clear that his voice was reaching tens of millions of Americans.

Trump's campaign renewed my hopes for the country, and I was sure that, finally, at long last, working men and women and their families had a champion like no other. And when President Trump took on the endless wars of the Bush years and the Obama years, my hopes grew to near certainty that he would be elected president. He was clearly an American original, a great patriot, and fearless.

These are not qualities you run into often in the overly polite and sophisticated circles of the Washington and New York elites. There's seldom much difference between a Republican and a Democrat in that rarefied air. And the wealthier and more powerful the dinner table guests, the more likely you are to be in the presence of global elites who expect the United States to be the world's policeman, to run endless deficits to support endless wars.

The globalist elites are quite used to insisting on America Not First, sometimes America Not At All. One evening, my wife and I were at a dinner on the Upper East Side of Manhattan, and the guest of honor was a former four-star general. He regaled the table with his view of the United States' role in the world and why we must stay in Afghanistan.

After some time, much of it spent biting my lip, I questioned his version of events and what he saw as a necessity of giving up the precious lives of our young service members and trillions of dollars as the United States followed a strategy that apparently didn't consider victory as the necessary objective for US generals.

He further extolled the virtues of the Long War Doctrine, and I drew a few gasps from the other guests when I said, "General, forgive me, but the long war doesn't appear to be a doctrine. It's more of a rationalization for a war without end, for a war that we haven't won, and may never win." For some reason the general took my statement as a personal offense and said, "In a different circumstance I would kick your ass."

At that moment, I was grateful for the current circumstances and replied, "You know, General, you may be right, but I'm used to suffering consequences for telling the truth." He rose from his chair and stormed away from the table, and I haven't seen him since; probably just as well.

For the better part of two decades, there wasn't much distinction among our leaders in their worldview. Whether they were military or corporate, finance, law, academia, or media, the globalist elites were part of a great gray orthodoxy that was impervious to challenge or change. Until Trump was elected. And hallelujah!

The president's battles with the radical Dems, the Deep State, and RINOs of the last four years would have been exhausting for most of us. And then there are the adversaries outside our borders, all around the world. He has met every challenge, every threat, yet he is far from exhausted. He is exultant and appears stronger than ever. Though we have no right to ask this president to take on another term, I believe we have every reason to do so.

President Trump has become the indomitable American patriot, he is the hero of our time. He stands with every American, every family, and he has never backed off and never backed down. There is simply no one I'd rather have lead the nation than President Trump. He became a historic president in his first term. I can't wait to see what he will do in his second!

All we are asking of him is to build the United States' greatest economy not once but twice. You know he can.

All we ask of him is that he keeps winning and makes more beautiful history for generations of Americans to come. You know he will.

ACKNOWLEDGMENTS

First and always, thanks to my wife, Debi, for her love and support throughout four decades of love and mostly laughter, and our wonderful family, four generations of us and more on the way.

My thanks as well to all my friends and colleagues at Fox Business, President Lauren Petterson, SVP Gary Schrier, Producers Jeff Field, Alex Hooper, Eric Schaffer, Chief Booker, Anne McCarton, writer/producers Bob Regan, Michael Biondi, Bria Stone, Andrew Espitallier, and John Fawcett.

My gratitude to Harper Executive Editor and VP/Editorial Director of Broadside Books Eric Nelson, and to my good friend and agent Wayne Kabak. Thanks to Dennis Kneale for his skill and getting the book over the finish line, and to John Carney for his many contributions.

This book is dedicated to President Trump, First Lady Melania Trump, and the entire Trump family for their many sacrifices and great courage in service to the nation.

ABOUT THE AUTHOR

LOU DOBBS is the *New York Times* bestselling author of six books and the host of the number one news program on business television, Lou Dobbs Tonight, on the Fox Business Network. He is also the host of the nationally syndicated Lou Dobbs Financial Reports, airing on the radio daily. Named "TV's Premier Business News Anchorman" by the *Wall Street Journal*, Dobbs has numerous Emmys, a CableACE Award, a Peabody Award, and many other distinguished honors.

DENNIS KNEALE has over thirty years of experience in journalism, from *Forbes* and the *Wall Street Journal* to anchor for CNBC and Fox Business. He is the CEO of Dennis Kneale Media, providing content, training, strategy, and investment advice.